An Introduction to
Tensor Calculus,
Relativity and Cosmology

An Introduction to Tensor Calculus, Relativity and Cosmology

Third Edition

D. F. Lawden
Department of Mathematics
The University of Aston in Birmingham

JOHN WILEY & SONS
Chichester · New York · Brisbane · Toronto · Singapore

First published 1962
by Methuen & Co Ltd
Second edition 1968
Reprinted twice
First published as a Science Paperback 1967
by Chapman and Hall Ltd
11 New Fetter Lane, London EC4P 4EE
Reprinted 1975

© 1962, 1967, 1975 *D. F. Lawden*

Copyright © 1982 by John Wiley & Sons Ltd.

Reprinted June 1986

Library of Congress Cataloging in Publication Data:

Lawden, Derek F.
 An introduction to tensor calculus, relativity and cosmology.—3rd ed.

 Rev. ed. of: An introduction to tensor calculus and relativity. 2nd ed.
 Bibliography: p.
 Includes index.
 1. Relativity (Physics) 2. Calculus of tensors. 3. Cosmology. I. Title.
QC173.55.L38   1982   530.1′1   81–14801
ISBN 0 471 10082 X (cloth)   AACR2
ISBN 0 471 10096 X (paper)

British Library Cataloguing in Publication Data:

Lawden, D. F.
 An introduction to tensor calculus, relativity
 and cosmology.—3rd ed.
 1. Relativity (Physics)
 2. Calculus of tensors
 3. Cosmology—Mathematics
 I. Title
 530.1′1   QC173.55

ISBN 0 471 10082 X (cloth)
ISBN 0 471 10096 X (paper)

Typeset by Macmillan India Ltd, Bangalore, and printed in the United States of America,
by Vail-Ballou Press Inc., Binghamton, N.Y.

Contents

Preface

The revolt against the ancient world view of a universe centred upon the earth, which was initiated by Copernicus and further developed by Kepler, Galileo and Newton, reached its natural termination in Einstein's theories of relativity. Starting from the concept that there exists a unique privileged observer of the cosmos, namely man himself, natural philosophy has journeyed to the opposite pole and now accepts as a fundamental principle that all observers are equivalent, in the sense that each can explain the behaviour of the cosmos by application of the same set of natural laws. Another line of thought whose complete development takes place within the context of special relativity is that pioneered by Maxwell, electromagnetic field theory. Indeed, since the Lorentz transformation equations upon which the special theory is based constitute none other than the transformation group under which Maxwell's equations remain of invariant form, the relativistic expression of these equations discovered by Minkowski is more natural than Maxwell's. In the history of natural philosophy, therefore, relativity theory represents the culmination of three centuries of mathematical modelling of the macroscopic physical world; it stands at the end of an era and is a magnificent and fitting memorial to the golden age of mathematical physics which came to an end at the time of the First World War. Einstein's triumph was also his tragedy; although he was inspired to create a masterpiece, this proved to be a monument to the past and its very perfection a barrier to future development. Thus, although all the implications of the general theory have not yet been uncovered, the barrenness of Einstein's later explorations indicates that the growth areas of mathematical physics lie elsewhere, presumably in the fecund soil of quantum and elementary-particle theory.

Nevertheless, relativity theory, especially the special form, provides a foundation upon which all later developments have been constructed and it seems destined to continue in this role for a long time yet. A thorough knowledge of its elements is accordingly a prerequisite for all students who wish to understand contemporary theories of the physical world and possibly to contribute to their expansion. This being universally recognized, university courses in applied mathematics and mathematical physics commonly include an introductory course in the subject at the undergraduate level, usually in the second and third years, but occasionally even in the first year. This book has been written to provide a suitable supporting text for such courses. The author has taught this type of class for the past twenty-five years and has become very familiar with the difficulties regularly experienced by students when they first study this subject;

the identification of these perplexities and their careful resolution has therefore been one of my main aims when preparing this account. To assist the student further in mastering the subject, I have collected together a large number of exercises and these will be found at the end of each chapter; most have been set as course work or in examinations for my own classes and, I think, cover almost all aspects normally treated at this level. It is hoped, therefore, that the book will also prove helpful to lecturers as a source of problems for setting in exercise classes.

When preparing my plan for the development of the subject, I decided to disregard completely the historical order of evolution of the ideas and to present these in the most natural logical and didactic manner possible. In the case of a fully established (and, indeed, venerable) theory, any other arrangement for an introductory text is unjustifiable. As a consequence, many facets of the subject which were at the centre of attention during the early years of its evolution have been relegated to the exercises or omitted entirely. For example, details of the seminal Michelson–Morley experiment and its associated calculations have not been included. Although this event was the spark which ignited the relativistic tinder, it is now apparent that this was an historical accident and that, being implicit in Maxwell's principles of electromagnetism, it was inevitable that the special theory would be formulated near the turn of the century. Neither is the experiment any longer to be regarded as a crucial test of the theory, since the theory's manifold implications for all branches of physics have provided countless other checks, all of which have told in its favour. The early controversies attending the birth of relativity theory are, however, of great human interest and students who wish to follow these are referred to the books by Clark, Hoffmann and Lanczos listed in the Bibliography at the end of this book.

A curious feature of the history of the special theory is the persistence of certain paradoxes which arose shortly after it was first propounded by Einstein and which were largely disposed of at that time. In spite of this, they are rediscovered every decade or so and editors of popular scientific periodicals (and occasionally, and more reprehensibly, serious research journals) seem happy to provide space in which these old battles can be refought, thus generating a good deal of acrimony on all sides (and, presumably, improving circulation). The source of the paradoxes is invariably a failure to appreciate that the special theory is restricted in its validity to inertial frames of reference or an inability to jettison the Newtonian concept of a unique ordering of events in time. Complete books based on these misconceptions have been published by authors who should know better, thus giving students the unfortunate impression that the consistency of this system of ideas is still in doubt. I have therefore felt it necessary to mention some of these 'paradoxes' at appropriate points in the text and to indicate how they are resolved; others have been used as a basis for exercises, providing excellent practice for the student to train himself to think relativistically.

Much of the text was originally published in 1962 under the title *An Introduction to Tensor Calculus and Relativity*. All these sections have been thoroughly revised in the light of my teaching experience, one or two sections have been discarded as containing material which has proved to be of little importance for an understanding of the basics (e.g. relative tensors) and a number of new sections have been added (e.g. equations of motion of an elastic fluid, black holes, gravitational waves, and a more detailed account of the relationship between the metric and affine connections). But the main improvement is the addition of a chapter covering the application of the general theory to cosmology. As a result of the great strides made in the development of optical and, particularly, radio astronomy during the last twenty years, cosmological science has moved towards the centre of interest for physics and very few university courses in the general theory now fail to include lectures in this area.

It is a common (and desirable) practice to provide separate courses in the special and general theories, the special being covered in the second or third under-graduate year and the general in the final year of the undergraduate course or the first year of a postgraduate course. The book has been arranged with this in mind and the first four chapters form a complete unit, suitable for reading by students who may not progress to the general theory. Such students need not be burdened with the general theory of tensors and Riemannian spaces, but can acquire a mastery of the principles of the special theory using only the unsophisticated tool of Cartesian tensors in Euclidean (or quasi-Euclidean) space. In my experience, even students who intend to take a course in the general theory also benefit from exposure to the special theory in this form, since it enables them to concentrate upon the difficulties of the relativity principles and not to be distracted by avoidable complexities of notation. I have no sympathy with the teacher who, encouraged by the shallow values of the times, regards it as a virtue that his lectures exhibit his own present mastery of the subject rather than his appreciation of his students' bewilderment on being led into unfamiliar territory. All students should, in any case, be aware of the simpler form the theory of tensors assumes when the transformation group is restricted to be orthogonal.

As a consequence of my decision to develop the special theory within the context of Cartesian tensors, it was necessary to reduce the special relativistic metric to Pythagorean form by the introduction of either purely imaginary spatial coordinates or a purely imaginary time coordinate for an event. I have followed Minkowski and put $x_4 = ict$; thus, the metric has necessarily been taken in the form

$$ds^2 = dx_1{}^2 + dx_2{}^2 + dx_3{}^2 + dx_4{}^2 = dx^2 + dy^2 + dz^2 - c^2 dt^2$$

and ds has the dimension of length. I have retained this definition of the interval between two events observed from a freely falling frame in the general theory; this not only avoids confusion but, in the weak-field approximation, permits the distinction between covariant and contravariant components of a tensor to be eliminated by the introduction of an imaginary time. A disadvantage is that ds is imaginary for timelike intervals and the interval parameter s accordingly takes imaginary values along the world-line of any material body. Thus, when writing down the equations for the geodesic world-line of a freely falling body, it is usually convenient to replace s by τ, defined by the equation $s = ic\tau$, τ being called the proper time and $d\tau$ the proper time interval. However, it is understood throughout the exposition of the general theory that the metric tensor for space-time g_{ij} is such that $ds^2 = g_{ij}\, dx^i\, dx^j$; a consequence is that the cosmical constant term in Einstein's equation of gravitation has a sign opposite to that taken by some authors.

References in the text are made by author and year and have been collected together at the end of the book.

D. F. LAWDEN

Department of Mathematics,
The University of Aston in Birmingham.
May, 1981.

List of Constants

In the SI system of units:

Gravitational constant $= G = 6.673 \times 10^{-11} \, \mathrm{m^3 \, kg^{-1} \, s^{-2}}$

Velocity of light $= c = 2.998 \times 10^8 \, \mathrm{ms^{-1}}$

Mass of sun $= 1.991 \times 10^{30} \, \mathrm{kg}$

Mass of earth $= 5.979 \times 10^{24} \, \mathrm{kg}$

Mean radius of earth $= 6.371 \times 10^6 \, \mathrm{m}$

Permittivity of free space $= \varepsilon_0 = 8.854 \times 10^{-12}$

Permeability of free space $= \mu_0 = 1.257 \times 10^{-6}$

CHAPTER 1

Special Principle of Relativity. Lorentz Transformations

1. Newton's laws of motion

A proper appreciation of the physical content of Newton's three laws of motion is an essential prerequisite for any study of the special theory of relativity. It will be shown that these laws are in accordance with the fundamental principle upon which the theory is based and thus they will also serve as a convenient introduction to this principle.

The first law states that *any particle which is not subjected to forces moves along a straight line at constant speed.* Since the motion of a particle can only be specified relative to some coordinate frame of reference, this statement has meaning only when the reference frame to be employed when observing the particle's motion has been indicated. Also, since the concept of force has not, at this point, received a definition, it will be necessary to explain how we are to judge when a particle is 'not subjected to forces'. It will be taken as an observed fact that if rectangular axes are taken with their origin at the centre of the sun and these axes do not rotate relative to the most distant objects known to astronomy, viz. the extragalactic nebulae, then the motions of the neighbouring stars relative to this frame are very nearly uniform. The departure from uniformity can reasonably be accounted for as due to the influence of the stars upon one another and the evidence available suggests very strongly that if the motion of a body in a region infinitely remote from all other bodies could be observed, then its motion would always prove to be uniform relative to our reference frame irrespective of the manner in which the motion was initiated.

We shall accordingly regard the first law as asserting that, in a region of space remote from all other matter and empty save for a single test particle, a reference frame can be defined relative to which the particle will always have a uniform motion. Such a frame will be referred to as an *inertial frame.* An example of such an inertial frame which is conveniently employed when discussing the motions of bodies within the solar system has been described above. However, if S is any inertial frame and $\bar{S}$ is another frame whose axes are always parallel to those of S but whose origin moves with a constant velocity $\mathbf{u}$ relative to S, then $\bar{S}$ also is

1

inertial. For, if $\mathbf{v}, \bar{\mathbf{v}}$ are the velocities of the test particle relative to $S, \bar{S}$ respectively, then

$$\bar{\mathbf{v}} = \mathbf{v} - \mathbf{u} \qquad (1.1)$$

and, since $\mathbf{v}$ is always constant, so is $\bar{\mathbf{v}}$. It follows, therefore, that a frame whose origin is at the earth's centre and whose axes do not rotate relative to the stars can, for most practical purposes, be looked upon as an inertial frame, for the motion of the earth relative to the sun is very nearly uniform over periods of time which are normally the subject of dynamical calculations. In fact, since the earth's rotation is slow by ordinary standards, a frame which is fixed in this body can also be treated as approximately inertial and this assumption will only lead to appreciable errors when motions over relatively long periods of time are being investigated, e.g. Foucault's pendulum, long-range gunnery calculations. A frame attached to a non-rotating spaceship, whose rocket motor is inoperative and which is moving in a negligible gravitational field (e.g. in interstellar space), provides another example of an inertial frame. Since the stars of our galaxy move uniformly relative to one another over very long periods of time, the frames attached to them will all be inertial provided they do not rotate relative to the other galaxies.

Having established an inertial frame, if it is found by observation that a particle does not have a uniform motion relative to the frame, the lack of uniformity is attributed to the action of a *force* which is exerted upon the particle by some agency. For example, the orbits of the planets are considered to be curved on account of the force of gravitational attraction exerted upon these bodies by the sun and when a beam of charged particles is observed to be deflected when a bar magnet is brought into the vicinity, this phenomenon is understood to be due to the magnetic forces which are supposed to act upon the particles. If $\mathbf{v}$ is the particle's velocity relative to the frame at any instant t, its acceleration $\mathbf{a} = d\mathbf{v}/dt$ will be non-zero if the particle's motion is not uniform and this quantity is accordingly a convenient measure of the applied force $\mathbf{f}$. We take, therefore,

$$\mathbf{f} \propto \mathbf{a}$$

or
$$\mathbf{f} = m\mathbf{a} \qquad (1.2)$$

where m is a constant of proportionality which depends upon the particle and is termed its *mass*. The definition of the mass of a particle will be given almost immediately when it arises quite naturally out of the third law of motion. Equation (1.2) is essentially a definition of force relative to an inertial frame and is referred to as the *second law of motion*. It is sometimes convenient to employ a non-inertial frame in dynamical calculations, in which case a body which is in uniform motion relative to an inertial frame and is therefore subject to no forces, will nonetheless have an acceleration in the non-inertial frame. By equation (1.2), to this acceleration there corresponds a force, but this will not be attributable to any obvious agency and is therefore usually referred to as a 'fictitious' force. Well-known examples of such forces are the centrifugal and Coriolis forces associated with frames which are in uniform rotation relative to an inertial frame, e.g. a frame rotating with the earth. By introducing such 'fictitious' forces, the second law of motion becomes applicable in all reference frames. Such forces are called *inertial forces* (see Section 44).

According to the third law of motion, *when two particles P and Q interact so as to influence one another's motion, the force exerted by P on Q is equal to that exerted by Q on P but is in the opposite sense.* Defining the *momentum* of a particle relative to a reference frame as the product of its mass and its velocity, it is proved in elementary textbooks that the second and third laws taken together imply that the sum of the momenta of any two particles involved in a collision is conserved. Thus, if m_1, m_2 are the masses of two such particles and $\mathbf{u}_1$, $\mathbf{u}_2$ are their respective velocities immediately before the collision and $\mathbf{v}_1$, $\mathbf{v}_2$ are their respective velocities immediately afterwards, then

$$m_1 \mathbf{u}_1 + m_2 \mathbf{u}_2 = m_1 \mathbf{v}_1 + m_2 \mathbf{v}_2 \tag{1.3}$$

i.e.

$$\frac{m_2}{m_1} (\mathbf{u}_2 - \mathbf{v}_2) = \mathbf{v}_1 - \mathbf{u}_1 \tag{1.4}$$

This last equation implies that the vectors $\mathbf{u}_2 - \mathbf{v}_2$, $\mathbf{v}_1 - \mathbf{u}_1$ are parallel, a result which has been checked experimentally and which constitutes the physical content of the third law. However, equation (1.4) shows that the third law is also, in part, a specification of how the mass of a particle is to be measured and hence provides a definition for this quantity. For

$$\frac{m_2}{m_1} = \frac{|\mathbf{v}_1 - \mathbf{u}_1|}{|\mathbf{u}_2 - \mathbf{v}_2|} \tag{1.5}$$

and hence the ratio of the masses of two particles can be found from the results of a collision experiment. If, then, one particular particle is chosen to have unit mass (e.g. the standard kilogramme), the masses of all other particles can, in principle, be determined by permitting them to collide with this standard and then employing equation (1.5).

2. Covariance of the laws of motion

It has been shown in the previous section that the second and third laws are essentially definitions of the physical quantities force and mass relative to a given reference frame. In this section, we shall examine whether these definitions lead to different results when different inertial frames are employed.

Consider first the definition of mass. If the collision between the particles m_1, m_2 is observed from the inertial frame $\overline{S}$, let $\overline{\mathbf{u}}_1$, $\overline{\mathbf{u}}_2$ be the particle velocities before the collision and $\overline{\mathbf{v}}_1$, $\overline{\mathbf{v}}_2$ the corresponding velocities after the collision. By equation (1.1),

$$\overline{\mathbf{u}}_1 = \mathbf{u}_1 - \mathbf{u}, \quad \text{etc.} \tag{2.1}$$

and hence

$$\bar{\mathbf{v}}_1 - \bar{\mathbf{u}}_1 = \mathbf{v}_1 - \mathbf{u}_1, \quad \bar{\mathbf{u}}_2 - \bar{\mathbf{v}}_2 = \mathbf{u}_2 - \mathbf{v}_2 \tag{2.2}$$

It follows that if the vectors $\mathbf{v}_1 - \mathbf{u}_1, \mathbf{u}_2 - \mathbf{v}_2$ are parallel, so are the vectors $\bar{\mathbf{v}}_1 - \bar{\mathbf{u}}_1$, $\bar{\mathbf{u}}_2 - \bar{\mathbf{v}}_2$ and consequently that, in so far as the third law is experimentally verifiable, it is valid in all inertial frames if it is valid in one. Now let $\bar{m}_1, \bar{m}_2$ be the particle masses as measured in $\bar{S}$. Then, by equation (1.5),

$$\frac{\bar{m}_2}{\bar{m}_1} = \frac{|\bar{\mathbf{v}}_1 - \bar{\mathbf{u}}_1|}{|\bar{\mathbf{u}}_2 - \bar{\mathbf{v}}_2|} = \frac{|\mathbf{v}_1 - \mathbf{u}_1|}{|\mathbf{u}_2 - \mathbf{v}_2|} = \frac{m_2}{m_1} \tag{2.3}$$

But, if the first particle is the unit standard, then $m_1 = \bar{m}_1 = 1$ and hence

$$\bar{m}_2 = m_2 \tag{2.4}$$

i.e. the mass of a particle has the same value in all inertial frames. We can express this by saying that mass is an *invariant* relative to transformations between inertial frames.

By differentiating equation (1.1) with respect to the time t, since $\mathbf{u}$ is constant it is found that

$$\bar{\mathbf{a}} = \mathbf{a} \tag{2.5}$$

where $\mathbf{a}, \bar{\mathbf{a}}$ are the accelerations of a particle relative to $S, \bar{S}$ respectively. Hence, by the second law (1.2), since $\bar{m} = m$, it follows that

$$\bar{\mathbf{f}} = \mathbf{f} \tag{2.6}$$

i.e. the force acting upon a particle is independent of the inertial frame in which it is measured.

It has therefore been shown that equations (1.2), (1.4) take precisely the same form in the two frames, $S, \bar{S}$, it being understood that mass, acceleration and force are independent of the frame and that velocity is transformed in accordance with equation (1.1). When equations preserve their form upon transformation from one reference frame to another, they are said to be *covariant* with respect to such a transformation. Newton's laws of motion are covariant with respect to a transformation between inertial frames.

3. Special principle of relativity

The special principle of relativity asserts that *all physical laws are covariant with respect to a transformation between inertial frames*. This implies that all observers moving uniformly relative to one another and employing inertial frames will be in agreement concerning the statement of physical laws. No such observer, therefore, can regard himself as being in a special relationship to the universe not shared by any other observer employing an inertial frame; there are no privileged observers. When man believed himself to be at the centre of creation both physically and spiritually, a principle such as that we have just enunciated would have been rejected as absurd. However, the revolution in attitude to our physical environment initiated by Copernicus has proceeded so far that today the principle is accepted as eminently reasonable and very strong evidence contradicting the principle would have to be discovered to disturb it as the foundation upon which theoretical physics is based. It is this principle which guarantees that observers inhabiting distant planets, belonging to stars whose motions may be very different from that of our own sun, will nevertheless be able to explain their local physical phenomena by application of the same physical laws we use ourselves.

It has been shown already that Newton's laws of motion obey the principle. Let us now transfer our attention to another set of fundamental laws governing non-mechanical phenomena, viz. Maxwell's laws of electrodynamics. These are more complex than the laws of Newton and are most conveniently expressed by the equations

$$\text{curl } \mathbf{E} = -\partial \mathbf{B}/\partial t \tag{3.1}$$

$$\text{curl } \mathbf{H} = \mathbf{j} + \partial \mathbf{D}/\partial t \tag{3.2}$$

$$\text{div } \mathbf{D} = \rho \tag{3.3}$$

$$\text{div } \mathbf{B} = 0 \tag{3.4}$$

where $\mathbf{E}$, $\mathbf{H}$ are the electric and magnetic intensities respectively, $\mathbf{D}$ is the displacement, $\mathbf{B}$ is the magnetic induction, $\mathbf{j}$ is the current density and ρ is the charge density (SI units are being used). Experiment confirms that these equations are valid when any inertial frame is employed. The most famous such experiment was that carried out by Michelson and Morley, who verified that the velocity of propagation of light waves in any direction is always measured to be c ($= 3 \times 10^8 \text{ m s}^{-1}$) relative to an apparatus stationary on the earth. As is well known, light has an electromagnetic character and this result is predicted by equations (3.1)–(3.4). However, the velocity of the earth in its orbit at any time differs from its velocity six months later by twice the orbital velocity, viz. 60 km/s and thus, by taking measurements of the velocity of light relative to the earth on two days separated by this period of time and showing them to be equal, it is possible to confirm that Maxwell's equations conform to the special principle of relativity. This is effectively what Michelson and Morley did. However, this interpretation of the results of their experiment was not accepted immediately, since it was thought that electromagnetic phenomena were supported by a medium called the *aether* and that Maxwell's equations would prove to be valid only in an inertial frame stationary in this medium, i.e. the special principle of relativity was denied for electromagnetic phenomena. It was supposed that an 'aether wind' would blow through an inertial frame not at rest in the aether and that this would have a disturbing effect on the propagation of electromagnetic disturbances through the medium, in the same way that a wind in the atmosphere affects the spread of sound waves. In such a frame, Maxwell's equations would (it was surmised) need correction by the inclusion of terms involving the wind velocity. That this would imply that terrestrial electrical machinery would behave differently in winter and summer does not appear to have raised any doubts!

After Michelson and Morley's experiment, a long controversy ensued and, though this is of great historical interest, it will not be recounted in this book. The special principle is now firmly established and is accepted on the grounds that the conclusions which may be deduced from it are everywhere found to be in conformity with experiment and also because it is felt to possess *a priori* a high degree of plausibility. A description of the steps by which it ultimately came to be appreciated that the principle was of quite general application would therefore be superfluous in an introductory text. It is, however, essential for our future development of the theory to understand the prime difficulty preventing an early acceptance of the idea that the electromagnetic laws are in conformity with the special principle.

Consider the two inertial frames S, $\overline{S}$. Suppose that an observer employing S measures the velocity of a light pulse and finds it to be $\mathbf{c}$. If the velocity of the same light pulse is measured by an observer employing the frame $\overline{S}$, let this be $\overline{\mathbf{c}}$. Then, by equation (1.1),

$$\overline{\mathbf{c}} = \mathbf{c} - \mathbf{u} \tag{3.5}$$

and it is clear that, in general, the magnitudes of the vectors $\overline{\mathbf{c}}$, $\mathbf{c}$ will be different. It appears to follow, therefore, that either Maxwell's equations (3.1)–(3.4) must be modified, or the special principle of relativity abandoned for electromagnetic phenomena. Attempts were made (e.g. by Ritz) to modify Maxwell's equations, but certain consequences of the modified equations could not be confirmed experimentally. Since the special principle was always found to be valid, the only remaining alternative was to reject equation (1.1) and to replace it by another in conformity with the experimental result that the speed of light is the same in all inertial frames. As will be shown in the next section, this can only be done at the expense of a radical revision of our intuitive ideas concerning the nature of space and time and this was very understandably strongly resisted.

4. Lorentz transformations. Minkowski space–time

The argument of this section will be founded on the following three postulates:

Postulate 1. A particle free to move under no forces has constant velocity in any inertial frame.

Postulate 2. The speed of light relative to any inertial frame is c in all directions.

Postulate 3. The geometry of space is Euclidean in any inertial frame.

Let the reference frame S comprise rectangular Cartesian axes $Oxyz$. We shall assume that the coordinates of a point relative to this frame are measured by the usual procedure and employing a measuring scale which is stationary in S (it is necessary to state this precaution, since it will be shown later that the length of a bar is not independent of its motion). It will also be supposed that standard atomic clocks, stationary relative to S, are distributed throughout space and are all synchronized with a master-clock at O. A satisfactory synchronization procedure would be as follows: Warn observers at all clocks that a light source at O will commence radiating at $t = t_0$. When an observer at a point P first receives light from this source, he is to set the clock at P to read $t_0 + OP/c$, i.e. it is assumed that light travels with a speed c relative to S, as found by experiment. The position and time of an event can now be specified relative to S by four coordinates (x, y, z, t), t being the time shown on the clock which is contiguous to the event. We shall often refer to the four numbers (x, y, z, t) as an *event*.

Let $\overline{O}\bar{x}\bar{y}\bar{z}$ be rectangular Cartesian axes determining the frame $\overline{S}$ (to be precise, these are rectangular as seen by an observer stationary in $\overline{S}$) and suppose that clocks at rest relative to this frame are synchronized with a master at $\overline{O}$. Any event can now be fixed relative to $\overline{S}$ by four coordinates $(\bar{x}, \bar{y}, \bar{z}, \bar{t})$, the space coordinates being measured by scales which are at rest in $\overline{S}$ and the time coordinate by the contiguous clock at rest in $\overline{S}$. If (x, y, z, t), $(\bar{x}, \bar{y}, \bar{z}, \bar{t})$ relate to the same event, in this section we are concerned to find the equations relating these corresponding coordinates. It is helpful to think of these transformation equations as a dictionary which enables us to translate a statement relating to any set of events from the S-language to the $\overline{S}$-language (or vice versa).

The possibility that the length of a scale and the rate of a clock might be affected by uniform motion relative to a reference frame was ignored in early physical theories. Velocity measurements were agreed to be dependent upon the reference frame, but lengths and time measurements were thought to be absolute. In relativity theory, as will appear, very few quantities are absolute, i.e. are independent of the frame in which the measuring instruments are at rest.

To comply with Postulate 1, we shall assume that each of the coordinates $(\bar{x}, \bar{y}, \bar{z}, \bar{t})$ is a linear function of the coordinates (x, y, z, t). The inverse relationship is then of the same type. A particle moving uniformly in S with velocity (v_x, v_y, v_z) will have space coordinates (x, y, z) such that

$$x = x_0 + v_x t, \quad y = y_0 + v_y t, \quad z = z_0 + v_z t \qquad (4.1)$$

If linear expressions in the coordinates $(\bar{x}, \bar{y}, \bar{z}, \bar{t})$ are now substituted for (x, y, z, t), it will be found on solving for $(\bar{x}, \bar{y}, \bar{z},)$ that these quantities are linear in $\bar{t}$ and hence that the particle's motion is uniform relative to $\overline{S}$. In fact, it may be proved that only a linear transformation can satisfy the Postulate 1.

Now suppose that at the instant $t = t_0$ a light source situated at the point P_0 (x_0, y_0, z_0) in S radiates a pulse of short duration. At any later instant t, the wavefront will occupy the sphere whose centre is P_0 and radius $c(t - t_0)$. This has equation

$$(x - x_0)^2 + (y - y_0)^2 + (z - z_0)^2 = c^2(t - t_0)^2 \qquad (4.2)$$

Let $(\bar{x}_0, \bar{y}_0, \bar{z}_0)$ be the coordinates of the light source as observed from $\overline{S}$ at the instant $\bar{t} = \bar{t}_0$ the short pulse is radiated. At any later instant $\bar{t}$, in accordance with Postulate 2, the wavefront must also appear from $\overline{S}$ to occupy a sphere of radius

$c(\bar{t} - \bar{t}_0)$ and centre $(\bar{x}_0, \bar{y}_0, \bar{z}_0)$. This has equation

$$(\bar{x} - \bar{x}_0)^2 + (\bar{y} - \bar{y}_0)^2 + (\bar{z} - \bar{z}_0)^2 = c^2(\bar{t} - \bar{t}_0)^2 \tag{4.3}$$

Equations (4.2), (4.3) describe the same set of events in languages appropriate to S, $\bar{S}$ respectively. It follows that the equations relating the coordinates (x, y, z, t), $(\bar{x}, \bar{y}, \bar{z}, \bar{t})$ must be so chosen that, upon substitution for the 'barred' quantities appearing in equation (4.3) the appropriate linear expressions in the 'unbarred' quantities, equation (4.2) results.

A mathematical device due to Minkowski will now be employed. We shall replace the time coordinate t of any event observed in S by a purely imaginary coordinate $x_4 = ict$ ($i = \sqrt{-1}$). The space coordinates (x, y, z) of the event will be replaced by (x_1, x_2, x_3) so that

$$x = x_1, \quad y = x_2, \quad z = x_3, \quad ict = x_4 \tag{4.4}$$

and any event is then determined by four coordinates $x_i (i = 1, 2, 3, 4)$. A similar transformation to coordinates $\bar{x}_i$ will be carried out in $\bar{S}$. Equations (4.2), (4.3) can then be written

$$\sum_{i=1}^{4} (x_i - x_{i0})^2 = 0 \tag{4.5}$$

$$\sum_{i=1}^{4} (\bar{x} - \bar{x}_{i0})^2 = 0 \tag{4.6}$$

The $\bar{x}_i$ are to be linear functions of the x_i and such as to transform equation (4.6) into equation (4.5) and hence such that

$$\sum_{i=1}^{4} (\bar{x}_i - \bar{x}_{i0})^2 \rightarrow k \sum_{i=1}^{4} (x_i - x_{i0})^2 \tag{4.7}$$

k can only depend upon the relative velocity of S and $\bar{S}$. It is reasonable to assume that the relationship between the two frames is a reciprocal one, so that, when the inverse transformation is made from S to $\bar{S}$, then

$$\sum_{i=1}^{4} (x_i - x_{i0})^2 \rightarrow k \sum_{i=1}^{4} (\bar{x}_i - \bar{x}_{i0})^2 \tag{4.8}$$

But the transformation followed by its inverse must leave any function of the coordinates $\bar{x}_i$ unaltered and hence $k^2 = 1$. In the limit, as the relative motion of S and $\bar{S}$ is reduced to zero, it is clear that $k \rightarrow +1$. Hence $k \neq -1$ and we conclude that k is identically unity.

The x_i will now be interpreted as rectangular Cartesian coordinates in a four-dimensional Euclidean space which we shall refer to as $\mathscr{E}_4$. This space is termed *Minkowski space–time*. The left-hand member of equation (4.5) is then the square of the 'distance' between two points having coordinates x_i, x_{i0}. It is now clear that the $\bar{x}_i$ can be interpreted as the coordinates of the point x_i referred to some other rectangular Cartesian axes in $\mathscr{E}_4$. For such an interpretation will certainly enable us to satisfy the requirement (4.7) (with $k = 1$). Also, the x_i, $\bar{x}_i$ will then be related by equations of the form

$$\bar{x}_i = \sum_{j=1}^{4} a_{ij}x_j + b_i \qquad (4.9)$$

where $i = 1, 2, 3, 4$ and the a_{ij}, b_i are constants and this relationship is linear. The b_i are the coordinates of the origin of the first set of rectangular axes relative to the second set. The a_{ij} will be shown to satisfy certain identities in Chapter 2 (equations (8.14), (8.15)). It is proved in algebra texts that the relationship between the x_i and $\bar{x}_i$ must be of the form we are assuming, if it is (i) linear and (ii) such as to satisfy the requirement (4.7).

Changing back from the x_i, $\bar{x}_i$ to the original coordinates of an event by equations (4.4), the equations (4.9) provide a means of relating space and time measurements in S with the corresponding measurements in $\bar{S}$. Subject to certain provisos (e.g. an event which has real coordinates in S, must have real coordinates in $\bar{S}$), this transformation will be referred to as the *general Lorentz transformation*.

5. The special Lorentz transformation

We shall now investigate the special Lorentz transformation obtained by supposing that the $\bar{x}_i$-axes in $\mathscr{E}_4$ are obtained from the x_i-axes by a rotation through an angle α parallel to the $x_1 x_4$-plane. The origin and the x_2, x_3-axes are unaffected by the rotation and it will be clear after consideration of Fig. 1 therefore that

$$\begin{aligned}\bar{x}_1 &= x_1 \cos\alpha + x_4 \sin\alpha & \bar{x}_2 &= x_2 \\ \bar{x}_4 &= -x_1 \sin\alpha + x_4 \cos\alpha & \bar{x}_3 &= x_3 \end{aligned} \right\} \qquad (5.1)$$

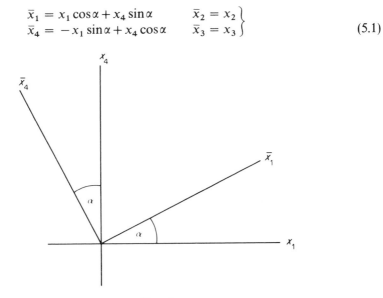

FIG. 1

10

Employing equations (4.4), these transformation equations may be written

$$\bar{x} = x\cos\alpha + ict\sin\alpha \qquad \bar{y} = y$$
$$ic\bar{t} = -x\sin\alpha + ict\cos\alpha \qquad \bar{z} = z \tag{5.2}$$

To interpret the equations (5.2), consider a plane which is stationary relative to the $\bar{S}$ frame and has equation

$$\bar{a}\bar{x} + \bar{b}\bar{y} + \bar{c}\bar{z} + \bar{d} = 0 \tag{5.3}$$

for all $\bar{t}$. Its equation relative to the S frame will be

$$(\bar{a}\cos\alpha)x + \bar{b}y + \bar{c}z + \bar{d} + ict\bar{a}\sin\alpha = 0 \tag{5.4}$$

at any fixed instant t. In particular, if $\bar{a} = \bar{b} = \bar{d} = 0$, this is the coordinate plane $\overline{Oxy}$ and its equation relative to S is $z = 0$, i.e. it is the plane Oxy. Again, if $\bar{b} = \bar{c} = \bar{d} = 0$, the plane is $\overline{Oyz}$ and its equation in S is

$$x = -ict\tan\alpha \tag{5.5}$$

i.e. it is a plane parallel to Oyz displaced a distance $-ict\tan\alpha$ along Ox. Finally, if $\bar{a} = \bar{c} = \bar{d} = 0$, the plane is $\overline{Ozx}$ and its equation with respect to S is $y = 0$, i.e. it is the plane Ozx. We conclude, therefore, that the Lorentz transformation equations (5.2) correspond to the particular case when the coordinate planes comprising $\bar{S}$ are obtained from those comprising S at any instant t by a translation along Ox a distance $-ict\tan\alpha$ (Fig. 2). Thus, if u is the speed of translation of $\bar{S}$ relative to S,

$$u = -ic\tan\alpha \tag{5.6}$$

It should also be noted that the events

$$x = y = z = t = 0, \quad \bar{x} = \bar{y} = \bar{z} = \bar{t} = 0$$

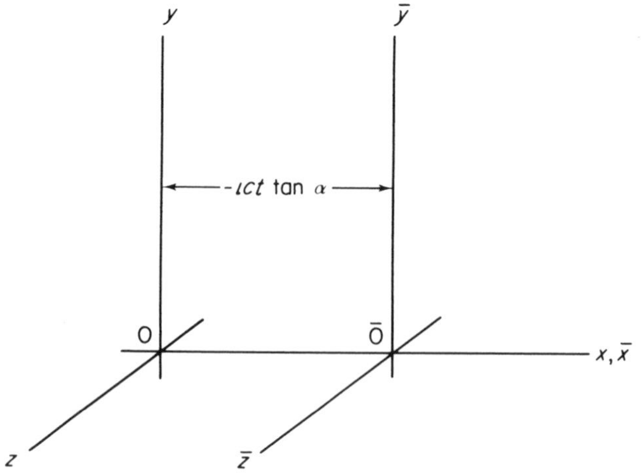

FIG. 2

correspond and hence that, at the instant O and $\overline{O}$ coincide, the S and $\overline{S}$ clocks at these points are supposed set to have zero readings; all other clocks are then synchronized with these.

Equation (5.6) indicates that α is imaginary and is directly related to the speed of translation. We have $\tan \alpha = iu/c$ and hence

$$\cos \alpha = \frac{1}{\sqrt{(1 - u^2/c^2)}}, \quad \sin \alpha = \frac{(iu/c)}{\sqrt{(1 - u^2/c^2)}} \tag{5.7}$$

Substituting in the equations (5.2), the special Lorentz transformation is obtained in its final form, viz.

$$\left. \begin{array}{ll} \overline{x} = \beta(x - ut) & \overline{y} = y \\ \overline{t} = \beta(t - ux/c^2) & \overline{z} = z \end{array} \right\} \tag{5.8}$$

where $\beta = (1 - u^2/c^2)^{-1/2}$.

If u is small by comparison with c, as is generally the case, these equations may evidently be approximated by the equations

$$\left. \begin{array}{ll} \overline{x} = x - ut & \overline{y} = y \\ \overline{t} = t & \overline{z} = z \end{array} \right\} \tag{5.9}$$

This set of equations, called the *special Galilean transformation* equations, is, of course, the set which was assumed to relate space and time measurements in the two frames in classical physical theory. However, the equation $\overline{t} = t$ was rarely stated explicitly, since it was taken as self-evident that time measurements were absolute, i.e. quite independent of the observer. It appears from equations (5.8) that this view of the nature of time can no longer be maintained and that, in fact, time and space measurements are related, as is shown by the dependence of $\overline{t}$ upon both t and x. This revolutionary idea is also suggested by the manner in which the special Lorentz transformation has been derived, viz. by a rotation of axes in a manifold which has both spacelike and timelike characteristics. However, this does not imply that space and time are now to be regarded as basically similar physical quantities, for it has only been possible to place the time coordinate on the same footing as the space coordinates in $\mathscr{E}_4$ by multiplying the former by i. Since x_4 must always be imaginary, whereas x_1, x_2, x_3 are real, the fundamentally different nature of space and time measurements is still maintained in the new theory.

If $u > c$, both $\overline{x}$ and $\overline{t}$ as given by equations (5.8) are imaginary. We conclude that no observer can possess a velocity greater than that of light relative to any other observer.

If equations (5.8) are solved for (x, y, z, t) in terms of $(\overline{x}, \overline{y}, \overline{z}, \overline{t})$, it will be found that the inverse transformation is identical with the original transformation, except that the sign of u is reversed. This also follows from the fact that the inverse transformation corresponds to a rotation of axes through an angle $-\alpha$ in space–time. Thus, the frame S has velocity $-\mathbf{u}$ when observed from $\overline{S}$.

6. Fitzgerald contraction. Time dilation

In the next two sections, we shall explore some of the more elementary physical consequences of the transformation equations (5.8).

Consider first a rigid rod stationary in $\overline{S}$ and lying along the $\overline{x}$-axis. Let $\overline{x} = \overline{x}_1, \overline{x} = \overline{x}_2$ at the two ends of the bar so that its length as measured in $\overline{S}$ is given by

$$\overline{l} = \overline{x}_2 - \overline{x}_1 \qquad (6.1)$$

In the frame S, the bar is moving with speed u and, to measure its length, it is necessary to observe the positions of its two ends at the same instant t. Suppose chalk marks are made on the x-axis at $x = x_1, x = x_2$, opposite the two ends, at the instant t. The making of these marks constitutes a pair of events with space–time coordinates $(x_1, t), (x_2, t)$ in S. In $\overline{S}$, this pair of events must have coordinates $(\overline{x}_1, \overline{t}_1), (\overline{x}_2, \overline{t}_2)$. Equations (5.8) now require that

$$\overline{x}_1 = \beta(x_1 - ut), \quad \overline{x}_2 = \beta(x_2 - ut) \qquad (6.2)$$

But $x_2 - x_1 = l$ is the length of the bar as measured in S and it follows by subtraction of equations (6.2) that

$$l = \overline{l}\sqrt{(1 - u^2/c^2)} \qquad (6.3)$$

The length of a bar accordingly suffers contraction when it is moved longitudinally relative to an inertial frame. This is the *Fitzgerald contraction*.

This contraction is not to be thought of as the physical reaction of the rod to its motion and as belonging to the same category of physical effects as the contraction of a metal rod when it is cooled. It is due to a changed relationship between the rod and the instruments measuring its length. $\overline{l}$ is a measurement carried out by scales which are stationary relative to the bar, whereas l is the result of a measuring operation with scales which are moving with respect to the bar. Also, the first operation can be carried out without the assistance of a clock, but the second operation involves simultaneous observation of the two ends of the bar and hence clocks must be employed. In classical physics, it was assumed that these two measurement procedures would yield the same result, since it was supposed that a rigid bar possessed intrinsically an attribute called its length and that this could in no way be affected by the procedure employed to measure it. It is now understood that length, like every other physical quantity, is *defined* by the procedure employed for its measurement and that it possesses no meaning apart from being the result of this procedure. From this point of view, it is not surprising that, when the procedure must be altered to suit the circumstances, the result will also be changed. It may assist the reader to adopt the modern view of the Fitzgerald contraction if we remark that the length of the rod considered above can be altered at any instant by simply changing our minds and commencing to employ the S frame rather than the $\overline{S}$ frame. Clearly, such a change of mathematical description can have no physical consequences.

Consider again the two events when chalk marks are made on the x-axis. Applying equations (5.8) to the space–time coordinates of the events in the two frames, the following equations are obtained:

$$\bar{t}_1 = \beta(t - ux_1/c^2), \quad \bar{t}_2 = \beta(t - ux_2/c^2) \tag{6.4}$$

These equations show that $\bar{t}_1 \neq \bar{t}_2$; i.e. although the events are simultaneous in S, they are not simultaneous in $\bar{S}$. The concept of simultaneity is accordingly also a relative one and has no absolute meaning as was previously thought.

The registration by the clock moving with $\bar{O}$ of the times $\bar{t}_1, \bar{t}_2$ constitutes two events having coordinates $(0, 0, 0, \bar{t}_1)$, $(0, 0, 0, \bar{t}_2)$ respectively in $\bar{S}$. Employing the inverse transformation to (5.8), it follows that the times t_1, t_2 of these events as measured in S are given by

$$t_1 = \beta \bar{t}_1, \qquad t_2 = \beta \bar{t}_2 \tag{6.5}$$

and hence that

$$\bar{t}_1 - \bar{t}_2 = (t_1 - t_2)\sqrt{(1 - u^2/c^2)} \tag{6.6}$$

This equation shows that the clock moving with $\bar{O}$ will appear from S to have its rate reduced by a factor $\sqrt{(1 - u^2/c^2)}$. This is the *time dilation* effect.

Since any physical process can be employed as a clock, the result just obtained implies that all physical processes will evolve more slowly when observed from a frame relative to which they are moving. Thus, the rate of decay of radioactive particles present in cosmic rays and moving with high velocities relative to the earth has been observed to be reduced by exactly the factor predicted by equation (6.6).

It may also be deduced that, if a human passenger were to be launched from the earth in a rocket which attained a speed approaching that of light and after proceeding to a great distance returned to the earth with the same high speed, suitable observations made from the earth would indicate that all physical processes occurring within the rocket, including the metabolic and physiological processes taking place inside the passenger's body, would suffer a retardation. Since all physical processes would be affected equally, the passenger would be unaware of this effect. Nonetheless, upon return to the earth he would find that his estimate of the duration of the flight was less than the terrestrial estimate. It may be objected that the passenger is entitled to regard himself as having been at rest and the earth as having suffered the displacement and therefore that the terrestrial estimate should be less than his own. This is the *clock paradox*. The paradox is resolved by observing that a frame moving with the rocket is subject to an acceleration relative to an inertial frame and consequently cannot be treated as inertial. The results of special relativity only apply to inertial frames and the rocket passenger is accordingly not entitled to make use of them in his own frame. As will be shown later, the methods of general relativity theory are applicable in any frame and it may be proved that, if the passenger employs these methods, his calculations will yield results in agreement with those obtained by the terrestrial observer.

Another clock paradox which requires more thought to resolve, can be stated thus: The clock at $\overline{O}$ runs slow when compared with O. But the frame $\overline{S}$ may be taken as the rest frame and then a similar argument proves that the moving clock O goes slow when compared with $\overline{O}$. This is a contradiction. Only inertial frames are involved, so that the paradox cannot be disposed of by rejecting one of the two calculations. Instead, it must be appreciated that a direct comparison of two clocks at different points in space cannot be made; the statement '$\overline{O}$ runs slow by comparison with O' needs amplification. The meaning special relativity theory gives to this sentence is: $\overline{O}$ is found to run slow when it is compared with the successive synchronized clocks, belonging to the frame S, with which it coincides during its motion. Similarly, 'O runs slow compared with $\overline{O}$' must be expanded to 'O runs slow when compared with the successive synchronized clocks belonging to $\overline{S}$ with which it coincides during its motion'. There is no contradiction between these expanded statements (see Exercise 19 at the end of this chapter).

7. Spacelike and timelike intervals. Light cone

We have proved in section 4 that if x_i, x_{i0} are the coordinates in Minkowski space–time of two events, then

$$\sum_{i=1}^{4} (x_i - x_{i0})^2 \tag{7.1}$$

is invariant, i.e. has the same value for all observers employing inertial frames and thus rectangular axes in space–time. Reverting by equations (4.4) to the ordinary space and time coordinates employed in an inertial frame, it follows that

$$(x - x_0)^2 + (y - y_0)^2 + (z - z_0)^2 - c^2(t - t_0)^2 \tag{7.2}$$

is invariant for all inertial observers.

Thus, if (x, y, z, t), (x_0, y_0, z_0, t_0) are the coordinates of two events relative to any inertial frame S and we define the *proper time interval* τ between the events by the equation

$$\tau^2 = (t - t_0)^2 - \frac{1}{c^2}\{(x - x_0)^2 + (y - y_0)^2 + (z - z_0)^2\} \tag{7.3}$$

then τ is an invariant for the two events. Two observers employing different inertial frames may attribute different coordinates to the events, but they will be in agreement concerning the value of τ.

Denoting the time interval between the events by Δt and the distance between them by Δd, both relative to the same frame S and positive, it follows from equation (7.3) that

$$\tau^2 = \Delta t^2 - \frac{1}{c^2}\Delta d^2 \tag{7.4}$$

Suppose that a new inertial frame $\overline{S}$ is now defined, moving in the direction of the line joining the events with speed $\Delta d/\Delta t$. This will only be possible if $\Delta d/\Delta t < c$.

Relative to this frame the events will occur at the same point and hence $\overline{\Delta d} = 0$. By equation (7.4), therefore,

$$\tau^2 = \overline{\Delta t}^2 \tag{7.5}$$

i.e. the proper time interval between two events is the ordinary time interval measured in a frame (if such exists) in which the events occur at the same space point. In this case, it is clear that $\tau^2 > 0$ and the proper time interval between the events is said to be *timelike*.

Suppose, if possible, that a frame $\overline{S}$ can be chosen relative to which the events are simultaneous. In this frame $\overline{\Delta t} = 0$ and

$$\tau^2 = -\frac{1}{c^2}\overline{\Delta d}^2 \tag{7.6}$$

Thus $\tau^2 < 0$, and, in any frame, $\Delta d/\Delta t > c$. τ is then purely imaginary and the interval between the events is said to be *spacelike*.

If the interval is timelike, $\Delta d/\Delta t < c$ and it is possible for a material body to be present at both events. On the other hand, if the interval is spacelike, $\Delta d/\Delta t > c$ and it is not possible for such a body to be present at both events. The intermediate case is when $\Delta d/\Delta t = c$ and then $\tau = 0$. Only a light pulse can be present at both events. It also follows that the proper time interval between the transmission and reception of a light signal is zero.

We shall now represent the event (x, y, z, t) by a point having these coordinates in a four-dimensional space. This space is also often referred to as Minkowski space–time but, unlike the space–time continuum introduced in section 4, it is not Euclidean. However, this representation has the advantage that the coordinates all take real values and it is therefore more satisfactory when diagrams are to be drawn. Suppose a particle is at the origin O of S at $t = 0$ and commences to move along Ox with constant speed u. Its y- and z-coordinates will always be zero and the representation of its motion in space–time will be confined to the xt-plane. In this plane, its motion will appear as the straight line QP, Q being the point $x = y = z = t = 0$ (Fig. 3). QP is called the *world-line* of the particle. If $\angle$ PQ$t = \theta$, $\tan \theta = u$. But $|u| \leqslant c$ and hence the world-line of the particle must lie in the sector AQB, where $\angle$ AQB $= 2\alpha$ and $\tan \alpha = c$. Similarly, the world-line of a particle which arrives at O at $t = 0$ after moving along Ox, must lie in the sector A'QB'. It follows that any event in either of these sectors must be separated from the event Q by a timelike interval, since a particle can be present at both events. Events in the sectors AQB', A'QB are separated from Q by spacelike intervals, since it is impossible for a particle to be present at such an event and also at Q. A'A, B'B are the world-lines of light signals passing through O at $t = 0$ and being propagated in the directions of the positive and negative x-axis respectively.

For any event in AQB, $t > 0$, i.e. it is in the future with respect to the event Q when the frame S is being employed. However, by no choice of frame can it be made simultaneous with Q, for this would imply a spacelike interval. *A fortiori*, in no frame can it occur prior to Q. The sector AQB accordingly contains events

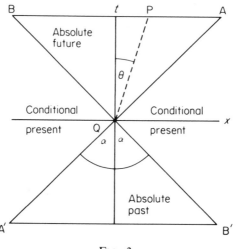

FIG. 3

which are in the *absolute future* with respect to the event Q. Similarly, all events in the sector A′ QB′ are in the *absolute past* with respect to Q. On the other hand, events lying in the sectors AQB′, A′QB are separated from Q by spacelike intervals and can all be made simultaneous with Q by proper choice of inertial frame. These events may occur before or after Q depending upon the frame being used. These two sectors define a region of space–time which will be termed the *conditional present*.

Since no physical signal can have a speed greater than c, the world-line of any such signal emanating from Q must lie in the sector AQB. It follows that the event Q can be the physical cause of only those events which are in the absolute future with respect to Q. Similarly, Q can be the effect of only those events in its absolute past. Q cannot be casually related to events in its conditional present.

This state of affairs should be contrasted with the essentially simpler situation of classical physics where there is no upper limit to the signal velocity and AA′, BB′ coincide along the x-axis. Past and future are then separated by a perfectly precise present in which events all have the time coordinate $t = 0$ for all observers.

In the four-dimensional space Q$xyzt$, the three regions of absolute past, absolute future and conditional present are separated from one another by the hyper-cone

$$x^2 + y^2 + z^2 - c^2t^2 = 0 \tag{7.7}$$

A light pulse transmitted from Q will have its world-line on this surface, which is accordingly called the *light cone* at Q. Since any arbitrary event can be selected to be Q, any event is the apex of a light cone which separates the space-time continuum in an absolute manner into three distinct regions relative to the event.

Exercises 1

1. A particle of mass m is moving in the plane of axes Oxy under the action of a force $\mathbf{f}$. Oxy is an inertial frame. $Ox'y'$ is rotating relative to the inertial frame so that $\angle\, x'Ox = \psi$ and $\dot\psi = \omega$. (r, θ) are polar coordinates of the particle relative to the rotating frame. If (f_r, f_θ) are the polar components of $\mathbf{f}$, (a_r, a_θ) are the polar components of the particle's acceleration relative to $Ox'y'$, v is the particle's speed relative to this frame and ϕ is the angle its direction of motion makes with the radius vector in this frame, obtain the equations of motion in the form

$$ma_r = f_r + 2m\omega v \sin\phi + mr\omega^2$$
$$ma_\theta = f_\theta - 2m\omega v \cos\phi - mr\dot\omega$$

Deduce that the motion relative to the rotating frame is in accordance with the second law if, in addition to $\mathbf{f}$, the following forces are also taken to act on the particle: (i) $m\omega^2 r$ radially outwards (the centrifugal force), (ii) $2m\omega v$ at right angles to the direction of motion (the Coriolis force), (iii) $mr\dot\omega$ transversely. (The latter force vanishes if the rotation is uniform.)

2. A bar lies along $\overline{O}\overline{x}$ and is stationary in $\overline{S}$. Show that if the positions of its ends are observed in S at instants which are simultaneous in $\overline{S}$, its length deduced from these observations will be greater than its length in $\overline{S}$ by a factor $(1 - u^2/c^2)^{-1/2}$.

3. Suppose that the bar referred to in Exercise 2 takes a time T to pass a fixed point on the x-axis, T being measured by a clock stationary at the fixed point. Defining the length of the bar in the S-frame to be uT, deduce the Fitzgerald contraction.

4. The measuring rod employed by S will appear from $\overline{S}$ to be shortened by a factor $(1 - u^2/c^2)^{1/2}$. Hence, when S measures the length of the bar fixed in $\overline{S}$ he might be expected to obtain the result

$$l = \overline{l}/(1 - u^2/c^2)^{1/2}$$

This contradicts equation (6.3). Resolve the contradiction. (*Hint*: It will be observed from $\overline{S}$ that S fixes the position of the forward end of the bar first and the position of the rear end a time $u\overline{l}/c^2$ later.)

5. A bar lies stationary along the x-axis of S. Show that the world-lines of the particles of the bar occupy a certain 'band' in the $x_1 x_4$-plane. By measuring the width of this 'band' parallel to the $\overline{x}_1$-axis, deduce the Fitzgerald contraction.

6. Verify that the Lorentz transformation equations (5.8) can be expressed in the form

$$\overline{x} = x\cosh\alpha - ct\sinh\alpha \qquad \overline{y} = y$$
$$c\overline{t} = ct\cosh\alpha - x\sinh\alpha \qquad \overline{z} = z$$

where $\tanh\alpha = u/c$. Deduce that

$$\overline{x} - c\overline{t} = (x - ct)e^\alpha, \qquad \overline{x} + c\overline{t} = (x + ct)e^{-\alpha}$$

Hence show that $x^2 - c^2t^2$ is invariant under this transformation. The clocks moving with the $\overline{S}$-frame are observed from the S-frame at the time t. Certain of them are seen to indicate the same time t. Show that these will lie in a plane relative to S and that this plane moves in S with velocity $c \tanh \frac{1}{2}\alpha$.

7. Two light pulses are moving in the positive direction along the x-axis of the frame S, the distance between them being d. Show that, as measured in $\overline{S}$, the distance between the pulses is

$$d\sqrt{\left(\frac{c+u}{c-u}\right)}$$

8. A and B are two points of an inertial frame S a distance d apart. An event occurs at B a time T (relative to clocks in S) after another event occurs at A. Relative to another inertial frame $\overline{S}$, the events are simultaneous. If **AP** is a displacement vector in S representing the velocity of $\overline{S}$ relative to S, prove that **P** lies in a plane perpendicular to AB, distance c^2T/d from A.

9. S, $\overline{S}$ are the inertial frames considered in section 5. The length of a moving rod, which remains parallel to the x and $\bar{x}$ axes, is measured as a in the frame S and $\bar{a}$ in the frame $\overline{S}$. By consideration of a Minkowski diagram for the rod, or otherwise, show that the rest length of the rod is

$$\frac{a\bar{a}\beta u/c}{\sqrt{(2\beta a\bar{a} - a^2 - \bar{a}^2)}}$$

10. If the position vectors $\mathbf{r} = (x, y, z)$, $\overline{\mathbf{r}} = (\bar{x}, \bar{y}, \bar{z})$ of an event as determined by the observers in the parallel inertial frames S, $\overline{S}$ respectively are mapped in the same independent $\mathscr{E}_3$, prove that

$$\overline{\mathbf{r}} = \mathbf{r} + \mathbf{u}\left\{\frac{\mathbf{u}\cdot\mathbf{r}}{u^2}(\beta - 1) + \beta t\right\}$$

$$\bar{t} = \beta(t + \mathbf{u}\cdot\mathbf{r}/c^2)$$

where $\mathbf{u}$ is the velocity of S as measured from $\overline{S}$.

11. A car 5 m long is to be placed in a garage only 3 m long. It is driven into the garage at four-fifths the speed of light c m/s by the owner; just before the bumpers strike the wall (which withstands the impact), show that the owner's wife can slam the doors. Calculate the length of the garage as seen by the driver and prove that he estimates that the car strikes the wall $4/c$ seconds before the doors are closed. Hence explain how the car fits into the garage from his point of view.

12. S, $\overline{S}$ are the two inertial frames related by the special Lorentz equations and u is the velocity of $\overline{S}$ relative to S. At $t = 0$ in S, particles A and B are at the points $(0, 0, 0)$ and $(d, 0, 0)$ respectively. Thereafter, both particles move along the x-axis with speed v a constant distance d apart. Write down equations describing the motion of the particles and, by transforming these to the frame $\overline{S}$ show that, in this frame, the particles are observed to move with a speed

$$\frac{v - u}{1 - uv/c^2}$$

a distance

$$\frac{d\,(1-u^2/c^2)^{1/2}}{1-uv/c^2}$$

apart.

13. S and $\bar{S}$ are the usual inertial frames having relative velocity u. A point moves along the $\bar{x}$-axis with constant acceleration c/τ starting from rest at $\bar{O}$ at $\bar{t}=0$. Write down its equation of motion in the frame $\bar{O}\bar{x}\bar{y}\bar{z}$. If $u=c/\sqrt{2}$, show that its motion in the frame $Oxyz$ is determined by the equation

$$x^2 - 2\sqrt{2}c\,(t+\tau)x + 2c^2 t(t+\tau) = 0$$

Deduce that, if t is small compared with τ, then

$$x = \frac{ct}{\sqrt{2}}(1+t/4\tau)$$

14. Two men are stationary in the S-frame at points on the x-axis separated by a distance d. They fire light pulses at one another simultaneously. In the $\bar{S}$-frame, show that one man A fires a time $\beta ud/c^2$ before the other man B and that, at the instant B fires, A's missile is still approaching B and is distant

$$d\left(\frac{c-u}{c+u}\right)^{1/2}$$

from him.

15. S and $\bar{S}$ are the usual pair of inertial frames having relative velocity u. The xz-plane of S is the surface of a lake. Waves are being propagated over this surface in the direction of the x-axis and are described by the equation $y = a\sin 2\pi f(t-x/w)$, f being the frequency and w the wave velocity. Obtain the equation which describes this wave motion in $\bar{S}$ and deduce that the frequency and wave velocity in this frame are given by

$$f' = f\beta(1-u/w), \quad w' = \frac{w-u}{1-uw/c^2}.$$

16. S, $\bar{S}$ are inertial frames. When observed from S, two events are simultaneous and at a distance D apart. When observed from $\bar{S}$, the time interval between the events is T. Calculate the distance between the events when observed from $\bar{S}$ and if, when viewed from S, the direction of motion of $\bar{S}$ relative to S makes an angle θ with the line joining the events, show that the relative speed of the frames is

$$c(1 + \frac{D^2}{c^2 T^2}\cos^2\theta)^{-1/2}.$$

17. Observed from a frame S, events A and B lie on the x-axis and B occurs a time T after A; the distance between the events is D. Calculate the velocity u of the

frame $\bar{S}$ relative to S if, observed from $\bar{S}$, the event B occurs a time T before A. What is the distance between the events as observed from $\bar{S}$? (Assume $D > cT$.) (Ans. $u = 2c^2 DT/(D^2 + c^2 T^2)$, $\bar{D} = D$.)

18. In the frame S, a particle is projected from O at $t = 0$ with the velocity components $(\frac{3}{4}c, \frac{1}{2}c, 0)$ and thereafter moves so that its acceleration is constant of magnitude g and is directed along the y-axis in the negative sense. Write down the x and y coordinates of the particle a time t later. The particle's motion is observed from the $\bar{S}$ frame. If $u = \frac{1}{2}c$, using the transformation equations (5.8), obtain equations for its coordinates $(\bar{x}, \bar{y})$ at time $\bar{t}$ in this frame. Deduce that the angle made by the velocity of projection with the $\bar{x}$-axis is $60°$ and that the particle's acceleration is $48g/25$.

19. The clocks at the origins of the frames S, $\bar{S}$ are synchronized to read zero when they pass one another and the clocks stationary in either frame are synchronized with the clock at the origin of the frame. At time t in S, the clock at $\bar{O}$ passes a clock C fixed to the x-axis of S. Show that C registers t and $\bar{O}$ registers $\sqrt{(1 - u^2/c^2)}t$ at this instant. Observed from $\bar{S}$, clock C is slowed by a factor $\sqrt{(1 - u^2/c^2)}$; in this frame, therefore, it might be expected that when C registers t, $\bar{O}$ would register $t/\sqrt{(1 - u^2/c^2)}$. Resolve the paradox by showing that, in the $\bar{S}$ frame, the clock C is not synchronized with O, but that C is always $u^2 t/c^2$ ahead of O.

20. In the frame S, at $t = 1$, a particle leaves the origin O and moves with constant velocity in the xy-plane having components $v_x = 5c/6$, $v_y = c/3$. What are the coordinates (x, y) of the particle at any later time t? If the velocity of $\bar{S}$ relative to S is $u = 3c/5$, calculate the coordinates $(\bar{x}, \bar{y})$ of the particle at time $\bar{t}$ in $\bar{S}$ and deduce that the particle makes its closest approach to $\bar{O}$ at time $\bar{t} = 220/113$.

21. $S, \bar{S}$ are the usual parallel frames with the origin $\bar{O}$ of $\bar{S}$ moving along the x-axis of S with velocity u. An observer A is stationed on the x-axis at $x = a$ and an observer $\bar{A}$ is stationed on the $\bar{x}$-axis at $\bar{x} = a$. Show that, in both the frames, the events (i) O passes $\bar{O}$ and (ii) A passes $\bar{A}$, are separated by a time T, but that the order of occurrence of these events is different. Calculate the value of T. If $T = a/3c$, show that $u = 3c/5$.

CHAPTER 2

Orthogonal Transformations. Cartesian Tensors

8. Orthogonal transformations

In section 4 events have been represented by points in a space $\mathscr{E}_4$. The resulting distribution of points was described in terms of their coordinates relative to a set of rectangular Cartesian axes. Each such set of axes was shown to correspond to an observer employing a rectangular Cartesian inertial frame in ordinary $\mathscr{E}_3$-space and clocks which are stationary in this frame. In this representation, the descriptions of physical phenomena given by two such inertial observers are related by a transformation in $\mathscr{E}_4$ from one set of rectangular axes to another. Such a transformation has been given at equation (4.9) and is called an *orthogonal* transformation. In general, if x_i, $\bar{x}_i (i = 1, 2, \ldots, N)$ are two sets of N quantities which are related by a linear transformation

$$\bar{x}_i = \sum_{j=1}^{N} a_{ij} x_j + b_i \qquad (8.1)$$

and, if the coefficients a_{ij} of this transformation are such that

$$\sum_{i=1}^{N} (\bar{x}_i - \bar{y}_i)^2 = \sum_{i=1}^{N} (x_i - y_i)^2 \qquad (8.2)$$

is an identity for all corresponding sets x_i, $\bar{x}_i$ and y_i, $\bar{y}_i$, then the transformation is said to be orthogonal. It is clear that the x_i, $\bar{x}_i$ may be thought of as the coordinates of a point in $\mathscr{E}_N$ referred to two different sets of rectangular Cartesian axes and then equation (8.2) states that the square of the distance between two points is an invariant, independent of the Cartesian frame.

Writing $z_i = x_i - y_i$, $\bar{z}_i = \bar{x}_i - \bar{y}_i$, it follows from equation (8.1) that

$$\bar{z}_i = \sum_{j=1}^{N} a_{ij} z_j \qquad (8.3)$$

Let z denote the column matrix with elements z_i, $\bar{z}$ the column matrix with elements $\bar{z}_i$ and A the $N \times N$ matrix with elements a_{ij}. Then the set of equations

21

(8.3) is equivalent to the matrix equation

$$\bar{z} = Az \tag{8.4}$$

Also, if z' is the transpose of z,

$$z'z = \sum_{i=1}^{N} z_i^2 \tag{8.5}$$

and thus the identity (8.2) may be written

$$\bar{z}'\bar{z} = z'z \tag{8.6}$$

But, from equation (8.4),

$$\bar{z}' = z'A' \tag{8.7}$$

Substituting in the left-hand member of equation (8.6) from equations (8.4), (8.7), it will be found that

$$z'A'Az = z'z \tag{8.8}$$

This can only be true for all z if

$$A'A = I \tag{8.9}$$

where I is the unit $N \times N$ matrix.

Taking determinants of both members of the matrix equation (8.9), we find that $|A|^2 = 1$ and hence

$$|A| = \pm 1 \tag{8.10}$$

A is accordingly regular. Let A^{-1} be its inverse. Multiplication on the right by A^{-1} of both members of equation (8.9) then yields

$$A' = A^{-1} \tag{8.11}$$

It now follows that

$$AA' = AA^{-1} = I \tag{8.12}$$

Let δ_{ij} be the ijth element of I, so that

$$\delta_{ij} \begin{rcases} = 1, & i = j \\ = 0, & i \neq j \end{rcases} \tag{8.13}$$

The symbols δ_{ij} are referred to as the *Kronecker deltas*. Equations (8.9), (8.12) are now seen to be equivalent to

$$\sum_{i=1}^{N} a_{ij}a_{ik} = \delta_{jk} \tag{8.14}$$

$$\sum_{i=1}^{N} a_{ji}a_{ki} = \delta_{jk} \tag{8.15}$$

respectively. These conditions are necessarily satisfied by the coefficients a_{ij} of the transformation (8.1) if it is orthogonal. Conversely, if either of these conditions is satisfied, it is easy to prove that equation (8.6) follows and hence that the transformation is orthogonal.

9. Repeated-index summation convention

At this point it is convenient to introduce a notation which will greatly abbreviate future manipulative work. It will be understood that, wherever in any term of an expression a literal index occurs twice, this term is to be summed over all possible values of the index. For example, we shall abbreviate by writing

$$\sum_{r=1}^{N} a_r b_r = a_r b_r \tag{9.1}$$

The index must be a literal one and we shall further stipulate that it must be a small letter. Thus $a_2 b_2$, $a_N b_N$ are individual terms of the expression $a_r b_r$, and no summation is intended in these cases.

Employing this convention, equations (8.14) and (8.15) can be written

$$a_{ij} a_{ik} = \delta_{jk}, \qquad a_{ji} a_{ki} = \delta_{jk} \tag{9.2}$$

respectively. Again, with $z_i = x_i - y_i$, equation (8.2) may be written

$$\bar{z}_i \bar{z}_i = z_i z_i \tag{9.3}$$

More than one index may be repeated in the same term, in which case more than one summation is intended. Thus

$$a_{ij} b_{jk} c_k = \sum_{j=1}^{N} \sum_{k=1}^{N} a_{ij} b_{jk} c_k \tag{9.4}$$

It is permissible to replace a repeated index by any other small letter, provided the replacement index does not occur elsewhere in the same term. Thus

$$a_i b_i = a_j b_j = a_k b_k \tag{9.5}$$

but

$$a_{ij} a_{ik} \neq a_{jj} a_{jk} \tag{9.6}$$

irrespective of whether the right-hand member is summed with respect to j or not. A repeated index shares this property with the variable of integration in a definite integral. Thus

$$\int_a^b f(x) \mathrm{d}x = \int_a^b f(y) \mathrm{d}y \tag{9.7}$$

A repeated index is accordingly referred to as a *dummy index*. Any other index will be called a *free index*. The same free index must appear in every term of an equation, but a dummy index may only appear in a single term.

The reader should note carefully the identity

$$\delta_{ij}a_j = a_i \tag{9.8}$$

for it will be of frequent application. δ_{ij} is often called a *substitution operator*, since when it multiplies a symbol such as a_j, its effect is to replace the index j by i.

10. Rectangular Cartesian tensors

Let x_i, y_i be rectangular Cartesian coordinates of two points Q, P respectively in $\mathscr{E}_N$. Writing $z_i = x_i - y_i$, the z_i are termed the *components* of the *displacement vector* **PQ** relative to the axes being used. If $\bar{x}_i$, $\bar{y}_i$ are the coordinates of Q, P with respect to another set of rectangular axes, the new coordinates will be related to the old by the transformation equations (8.1). Then, if $\bar{z}_i$ are the components of **PQ** in the new frame, it follows (equation (8.3)) that

$$\bar{z}_i = a_{ij}z_j \tag{10.1}$$

Any physical or geometrical quantity **A** having N components A_i defined in the x-frame and N components $\bar{A}_i$ similarly defined in the $\bar{x}$-frame, the two sets of components being related in the same manner as the components z_i, $\bar{z}_i$ of a displacement vector, i.e. such that

$$\bar{A}_i = a_{ij}A_j \tag{10.2}$$

is said to be a *vector* in $\mathscr{E}_N$ relative to rectangular Cartesian reference frames. We shall frequently abbreviate 'the vector whose components are A_i' to 'the vector A_i'.

If A_i, B_i are two vectors, consider the N^2 quantities A_iB_j. Upon transformation of axes, these quantities transform thus:

$$\bar{A}_i\bar{B}_j = a_{ik}a_{jl}A_kB_l \tag{10.3}$$

Any quantity having N^2 components C_{ij} defined in the x-frame and N^2 components $\bar{C}_{ij}$ defined similarly in the $\bar{x}$-frame, the two sets of components being related by a transformation equation

$$\bar{C}_{ij} = a_{ik}a_{jl}C_{kl} \tag{10.4}$$

is said to be a *tensor* of the second *rank*. We shall speak of 'the tensor C_{ij}'. Such a tensor is not, necessarily, representable as the product of two vectors.

A set of N^3 quantities D_{ijk} which transform in the same manner as the product of three vectors $A_iB_jC_k$, form a tensor of the third rank. The transformation law is

$$\bar{D}_{ijk} = a_{il}a_{jm}a_{kn}D_{lmn} \tag{10.5}$$

The generalization to a tensor of any rank should now be obvious. Vectors are, of course, tensors of the first rank.

If A_{ij}, B_{ij} are tensors, the sums $A_{ij} + B_{ij}$ are N^2 quantities which transform according to the same law as the A_{ij} and B_{ij}. The sum of two tensors of the second rank is accordingly also a tensor of this rank. This result can be generalized immediately to the sum of any two tensors of identical rank. Similarly, the difference of two tensors of the same rank is also a tensor.

Our method of introducing a tensor implies that the product of any number of vectors is a tensor. Quite generally, if $A_{ij} \ldots$, $B_{ij} \ldots$ are tensors of any ranks (which may be different), then the product $A_{ij} \ldots B_{kl} \ldots$ is a tensor whose rank is the sum of the ranks of the two factors. The reader should prove this formally for a product such as $A_{ij} B_{klm}$, by writing down the transformation equations. (N.B. the indices in the two factors must be kept distinct, for otherwise a summation is implied and this complicates matters; see section 12.)

The components of a tensor may be chosen arbitrarily relative to any one set of axes. The components of the tensor relative to any other set are then fixed by the transformation equations. Consider the tensor of the second rank whose components relative to the x_i-axes are the Kronecker deltas δ_{ij}. In the $\bar{x}_i$-frame, the components are

$$\bar{\delta}_{ij} = a_{ik} a_{jl} \delta_{kl} = a_{ik} a_{jk} = \delta_{ij} \tag{10.6}$$

by equations (9.2). Thus this tensor has the same components relative to all sets of axes. It is termed the *fundamental tensor* of the second rank.

If, to take the particular case of a third rank tensor as an example,

$$A_{ijk} = A_{jik} \tag{10.7}$$

for all values of i, j, k, A_{ijk} is said to be *symmetric* with respect to its indices i, j. Symmetry may be with respect to any pair of indices. If A_{ijk} is a tensor, its property of symmetry with respect to two indices is preserved upon transformation, for

$$\begin{aligned} \bar{A}_{jik} &= a_{jl} a_{im} a_{kn} A_{lmn} \\ &= a_{im} a_{jl} a_{kn} A_{mln} \\ &= \bar{A}_{ijk} \end{aligned} \tag{10.8}$$

where, in the second line, we have rearranged and put $A_{lmn} = A_{mln}$. Unless a property is preserved upon transformation, it will be of little importance to us, for we shall later employ tensors to express relationships which are valid for all observers and a chance relationship, true in one frame alone, will be of no fundamental significance.

Similarly, if

$$A_{ijk} = -A_{jik} \tag{10.9}$$

for all values of i, j, k, A_{ijk} is said to be *skew-symmetric* or *anti-symmetric* with respect to its first two indices. This property also is preserved upon transformation. Since $A_{11k} = -A_{11k}$, $A_{11k} = 0$. All components of A_{ijk} with the first two indices the same are clearly zero.

A tensor whose components are all zero in one frame, has zero components in every frame. A corollary to this result is that if $A_{ij} \ldots$, $B_{ij} \ldots$ are two tensors of the same rank whose corresponding components are equal in one frame, then they are equal in every frame. This follows because $A_{ij\ldots} - B_{ij\ldots}$ is a tensor whose components are all zero in the first frame and hence in every frame. Thus, a *tensor equation*

$$A_{ij\ldots} = B_{ij\ldots} \tag{10.10}$$

is valid for all choices of axes.

This explains the importance of tensors for our purpose. By expressing a physical law as a tensor equation in $\mathscr{E}_4$, we shall guarantee its covariance with respect to a change of inertial frame. A further advantage is that such an expression of the law also implies that it is covariant under a rotation and a translation of axes in $\mathscr{E}_3$, thus ensuring that the law conforms to the principles of *isotropy* and *homogeneity of space*.

The first principle states that all directions in space are equivalent in regard to the formulation of fundamental physical laws. Examples are that the inertia of a body in classical mechanics is independent of its direction of motion and that the power of attraction of an electric charge is the same in all directions. In the vicinity of the earth, the presence of the gravitational field tends to cloud our perception of the validity of this principle and the vertical direction at any point on the surface is sharply distinguished from any horizontal direction. But this is a purely local feature and the crew of a spaceship have no difficulty in accepting the principle. Mathematically, the principle requires that the equation expressing a basic physical law must not change its form when the reference frame is rotated. Laplace's equation $\nabla^2 V = 0$ is well known to possess this property, whereas the equation $\partial V/\partial x + \partial V/\partial y + \partial V/\partial z = 0$ does not; this explains why Laplace's equation occurs so frequently in mathematical physics, whereas the other equation does not.

The second principle affirms that all regions in space are also equivalent, i.e. that physical laws are the same in all parts of the cosmos. Covariance under a translation of axes is the mathematical expression of this requirement.

Both principles are almost certainly a consequence of the uniformity with which matter and radiation are distributed over the universe. It is doubtful whether either would be valid in a cosmos not possessing this property.

A less well-established third principle is that of *spatial parity*. This requires that physical laws should be impartial as between left- and right-handedness. The mathematical formulation of a law obeying this principle will be covariant under a transformation from a right-handed to a left-handed Cartesian frame or vice versa. Another way of expressing this principle is that, if the universe were observed in a mirror, the laws which would appear to govern its behaviour would be identical with the actual laws. For example, observation of the planetary motions in a mirror would alter their senses of rotation about the sun, but the law of gravitation would be unaffected. Although the more familiar laws are in conformity with this principle, those governing the behaviour of some fundamental particles do not appear to have this simple symmetry. Provided that the orthogonal transformations upon which our tensor calculus is being built are not restricted to be such that $|A| = +1$ (i.e. transformations between frames of opposite handedness are permitted), all tensor (and pseudotensor) equations will be in conformity with the principle of spatial parity.

From what has just been said, it is evident that the calculus of tensors is the natural language of mathematical physics, relativistic or non-relativistic. It guarantees that the equations being considered are of the type which can represent physical laws. However, in classical physics, a three-dimensional theory was adequate to ensure conformity with the principles of isotropy, homogeneity and parity of space. In relativistic physics, a four-dimensional theory is needed to incorporate the additional special principle of relativity.

11. Invariants. Gradients. Derivatives of tensors

Suppose that V is a quantity which is unaffected by any change of axes. Then V is called a *scalar invariant* or simply an *invariant*. Its transformation equation is simply

$$\bar{V} = V \tag{11.1}$$

As will be proved later (section 24), the charge of an electron is independent of the inertial frame from which it is measured and is, therefore, the type of quantity we are considering.

If a value of V is associated with each point of a region of $\mathscr{E}_N$, an *invariant field* is defined over this region. In this case V will be a function of the coordinates x_i. Upon transformation to new axes, V will be expressed in terms of the new coordinates $\bar{x}_i$; when so expressed, it is denoted by $\bar{V}$. Thus

$$\bar{V}(\bar{x}_1, \bar{x}_2, \ldots, \bar{x}_N) = V(x_1, x_2, \ldots, x_N) \tag{11.2}$$

is an identity. The reader should, perhaps, be warned that $\bar{V}$ is not, necessarily, the same function of the $\bar{x}_i$ that V is of the x_i.

If A_{ij} is a tensor, it is obvious that VA_{ij} is also a tensor of the second rank. It is therefore convenient to regard an invariant as a tensor of zero rank.

Consider the N partial derivatives $\partial V/\partial x_i$. These transform as a vector. To prove this it will be necessary to examine the transformation inverse to (8.1). In the matrix notation of section 8, this may be written

$$x = A^{-1}(\bar{x} - b) = A'(\bar{x} - b) \tag{11.3}$$

having made use of equation (8.11). Equation (11.3) is equivalent to

$$x_i = a'_{ij}(\bar{x}_j - b_j) \tag{11.4}$$

where a'_{ij} is the ijth element of A'. But $a'_{ij} = a_{ji}$ and hence

$$x_i = a_{ji}(\bar{x}_j - b_j) \tag{11.5}$$

It now follows that

$$\frac{\partial x_i}{\partial \overline{x}_j} = a_{ji} \tag{11.6}$$

and hence that

$$\frac{\partial \overline{V}}{\partial \overline{x}_i} = \frac{\partial V}{\partial x_j}\frac{\partial x_j}{\partial \overline{x}_i} = a_{ij}\frac{\partial V}{\partial x_j} \tag{11.7}$$

proving that $\partial V/\partial x_i$ is a vector. It is called the *gradient* of V and is denoted by grad V or ∇V.

If a tensor $A_{ij\ \ldots}$ is defined at every point of some region of $\mathcal{E}_N$, the result is a *tensor field*. The partial derivatives $\partial A_{ij\ \ldots}/\partial x_r$ can now be formed and constitute a tensor whose rank is one greater than that of $A_{ij\ \ldots}$. We shall prove this for a second rank tensor field A_{ij}. The argument is easily made general. We have

$$\begin{aligned}
\frac{\partial \overline{A}_{ij}}{\partial \overline{x}_k} &= \frac{\partial}{\partial \overline{x}_k}(a_{ir}a_{js}A_{rs})\\
&= \frac{\partial}{\partial x_t}(a_{ir}a_{js}A_{rs})\frac{\partial x_t}{\partial \overline{x}_k}\\
&= a_{ir}a_{js}a_{kt}\frac{\partial A_{rs}}{\partial x_t} \tag{11.8}
\end{aligned}$$

by equation (11.6).

12. Contraction. Scalar product. Divergence

If two indices are made identical, a summation is implied. Thus, consider A_{ijk}. Then

$$A_{ijj} = A_{i11} + A_{i22} + \ldots + A_{iNN} \tag{12.1}$$

There are N^3 quantities A_{ijk}. However, of the indices in A_{ijj}, only i remains free to range over the integers $1, 2, \ldots, N$, and hence there are but N quantities A_{ijj} and we could put $B_i = A_{ijj}$. The rank has been reduced by two and the process is accordingly referred to as *contraction*.

Contraction of a tensor yields another tensor. For example, if $B_i = A_{ijj}$ then, employing equations (9.2),

$$\overline{B}_i = \overline{A}_{ijj} = a_{iq}a_{jr}a_{js}A_{qrs} = a_{iq}\delta_{rs}A_{qrs} = a_{iq}A_{qrr} = a_{iq}B_q \tag{12.2}$$

Thus B_i is a vector. The argument is easily generalized.

In the special case of a tensor of rank two, e.g. A_{ij}, it follows that $\overline{A}_{ii} = A_{ii}$, i.e. A_{ii} is an invariant. Now, if A_i, B_i are vectors, A_iB_j is a tensor. Hence, A_iB_i is an invariant. This contracted product is called the *inner product* or the *scalar product* of the two vectors. We shall write

$$A_i B_i = \mathbf{A} \cdot \mathbf{B} \tag{12.3}$$

In particular, the scalar product of a vector with itself is an invariant. The positive square root of this invariant will be called the *magnitude* of the vector. Thus, if A is the magnitude of A_i, then

$$A^2 = A_i A_i = \mathbf{A} \cdot \mathbf{A} = \mathbf{A}^2 \tag{12.4}$$

In $\mathscr{E}_3$, if θ is the angle between two vectors $\mathbf{A}$ and $\mathbf{B}$, then

$$AB\cos\theta = \mathbf{A} \cdot \mathbf{B} \tag{12.5}$$

In $\mathscr{E}_N$, this equation is used to *define* θ. Hence, if

$$\mathbf{A} \cdot \mathbf{B} = 0 \tag{12.6}$$

then $\theta = \frac{1}{2}\pi$ and the vectors $\mathbf{A}$, $\mathbf{B}$ are said to be *orthogonal*.

If A_i is a vector field, $\partial A_i / \partial x_j$ is a tensor. By contraction it follows that $\partial A_i / \partial x_i$ is an invariant. This invariant is called the *divergence* of $\mathbf{A}$ and is denoted by div $\mathbf{A}$. Thus

$$\operatorname{div} \mathbf{A} = \frac{\partial A_i}{\partial x_i} \tag{12.7}$$

More generally, if $A_{ij\ldots}$ is a tensor field, $\partial A_{ij\ldots}/\partial x_r$ is a tensor. This tensor derivative can now be contracted with respect to the index r and any other index to yield another tensor, e.g. $\partial A_{ij\ldots}/\partial x_j$. This contraction is also referred to as the divergence of $A_{ij\ldots}$ with respect to the index j and we shall write

$$\frac{\partial A_{ij\ldots}}{\partial x_j} = \operatorname{div}_j A_{ij\ldots} \tag{12.8}$$

13. Pseudotensors

$\mathfrak{A}_{ij}$ is a pseudotensor if, when the coordinates are subjected to the transformation (8.1), its components transform according to the law

$$\overline{\mathfrak{A}}_{ij} = |A| a_{ik} a_{jl} \mathfrak{A}_{kl} \tag{13.1}$$

$|A|$ being the determinant of the transformation matrix A. Since for orthogonal transformations $|A| = \pm 1$ (equation (8.10)), relative to rectangular Cartesian frames, tensors and pseudotensors are identical except that, for certain changes of axes, all the components of a pseudotensor will be reversed in sign. For example, if in $\mathscr{E}_3$ a change is made from the right-handed system of axes to a left-handed system, the determinant of the transformation will be -1 and the components of a pseudotensor will then be subject to this additional sign change.

Let $e_{ij\ldots n}$ be a pseudotensor of the Nth rank which is skew-symmetric with respect to every pair of indices. Then all its components are zero, except those for which the indices $i, j, \ldots, n$ are all different and form a permutation of the numbers $1, 2, \ldots, N$. The effect of transposing any pair of indices in $e_{ij \ldots n}$ is to change its sign. It follows that if the arrangement $i, j, \ldots, n$ can be obtained from $1, 2, \ldots, N$ by an even number of transpositions, then $e_{ij \ldots n} = + e_{12 \ldots N}$, whereas if it can be obtained by an odd number $e_{ij \ldots n} = - e_{12 \ldots N}$. Relative to the x_i-axes let $e_{12 \ldots N} = 1$. Then, in this frame, $e_{ij \ldots n}$ is 0 if $i, j, \ldots, n$ is not a permutation of $1, 2, \ldots, N$, is $+1$ if it is an even permutation and is -1 if it is an odd permutation. Transforming to the $\bar{x}_i$-axes, we find that

$$\bar{e}_{12 \ldots N} = |A| a_{1i} a_{2j} \ldots a_{Nn} e_{ij \ldots n} = |A|^2 = 1 \tag{13.2}$$

But $\bar{e}_{ij \ldots n}$ is also skew-symmetric with respect to all its indices, since this is a property preserved by the transformation. Its components are also $0, \pm 1$ therefore and $e_{ij \ldots n}$ is a pseudotensor with the same components in all frames. It is called the *Levi–Civita* pseudotensor.

It may be shown without difficulty that:
(i) the sum or difference of two pseudotensors of the same rank is a pseudotensor.
(ii) the product of a tensor and a pseudotensor is a pseudotensor.
(iii) the product of two pseudotensors is a tensor.
(iv) the partial derivative of a pseudotensor with respect to x_i is a pseudotensor.
(v) a contracted pseudotensor is a pseudotensor.
Thus, to prove (iii), let $\mathfrak{A}_i, \mathfrak{B}_i$ be two pseudovectors. Then

$$\mathfrak{A}_i \mathfrak{B}_j = |A|^2 a_{ik} a_{jl} \mathfrak{A}_k \mathfrak{B}_l = a_{ik} a_{jl} \mathfrak{A}_k \mathfrak{B}_l \tag{13.3}$$

The method is clearly quite general. The remaining results will be left as exercises for the reader.

14. Vector products. Curl

Throughout this section we shall be assuming that $N = 3$, i.e. the space will be ordinary Euclidean space.

Let A_i, B_i be two vectors. Then $e_{ijk} A_r B_s$ is a pseudotensor of rank 5. Contracting twice, we get the pseudovector

$$\mathfrak{C}_i = e_{ijk} A_j B_k \tag{14.1}$$

whose components are

$$\left. \begin{array}{l} \mathfrak{C}_1 = A_2 B_3 - A_3 B_2 \\ \mathfrak{C}_2 = A_3 B_1 - A_1 B_3 \\ \mathfrak{C}_3 = A_1 B_2 - A_2 B_1 \end{array} \right\} \tag{14.2}$$

Provided we employ only right-handed systems of axes or only left-handed systems in $\mathscr{E}_3$, $\mathfrak{C}_i$ is indistinguishable from a vector. If, however, a change is made from a left-handed system to a right-handed system, or vice versa, the components of $\mathfrak{C}_i$ are multiplied by -1 in addition to the usual vector transformation. Since it is usual to employ only right-handed frames, $\mathfrak{C}_i$ is often referred to as a vector (or an *axial vector*) and treated as such. It is then called the *vector product* of **A** and **B** and we write

$$\mathfrak{C} = \mathbf{A} \times \mathbf{B} \tag{14.3}$$

Vector multiplication is non-commutative, for

$$\mathbf{B} \times \mathbf{A} = \mathfrak{e}_{ijk} B_j A_k = -\mathfrak{e}_{ikj} A_k B_j = -\mathbf{A} \times \mathbf{B} \tag{14.4}$$

having made use of $\mathfrak{e}_{ikj} = -\mathfrak{e}_{ijk}$. However, vector multiplication obeys the distributive law, for

$$\begin{aligned}
\mathbf{A} \times (\mathbf{B} + \mathbf{C}) &= \mathfrak{e}_{ijk} A_j (B_k + C_k) = \mathfrak{e}_{ijk} A_j B_k + \mathfrak{e}_{ijk} A_j C_k \\
&= \mathbf{A} \times \mathbf{B} + \mathbf{A} \times \mathbf{C} \tag{14.5}
\end{aligned}$$

We now introduce the abbreviated notation $\partial A_i / \partial x_j = A_{i,\,j}$. Any index after a comma will hereafter indicate a partial differentiation with respect to the corresponding coordinate; thus, $A_{i,\,jk}$ is a second derivative.

Now suppose A_i is a vector field. We can first construct a pseudotensor of rank 5 $\mathfrak{e}_{ijk} A_{r,\,s}$. Contracting twice, we get the pseudovector

$$\mathfrak{R}_i = \mathfrak{e}_{ijk} A_{k,\,j} \tag{14.6}$$

This has components

$$\begin{aligned}
\mathfrak{R}_1 &= \frac{\partial A_3}{\partial x_2} - \frac{\partial A_2}{\partial x_3} \\
\mathfrak{R}_2 &= \frac{\partial A_1}{\partial x_3} - \frac{\partial A_3}{\partial x_1} \\
\mathfrak{R}_3 &= \frac{\partial A_2}{\partial x_1} - \frac{\partial A_1}{\partial x_2}
\end{aligned} \Bigg\} \tag{14.7}$$

and is denoted by curl **A**. It, also, is an axial vector.

Equation (14.1) can still be employed to define a vector product when either or both of the vectors **A**, **B** are replaced by pseudotensors. If only one is replaced by a pseudovector the right-hand member of equation (14.1) will involve the product of two pseudovectors and a vector. The resulting vector product will then be a vector. Similarly, by replacing **A** in equation (14.6) by a pseudovector, the curl of a pseudovector is defined as an ordinary vector.

Exercises 2

1. Show that, in two dimensions, the general orthogonal transformation has matrix A given by

$$A = \begin{pmatrix} \cos\theta & \sin\theta \\ -\sin\theta & \cos\theta \end{pmatrix}$$

Verify that $|A| = 1$ and that $A^{-1} = A'$. T_{ij} is a tensor in this space. Write down in full the transformation equations for all its components and deduce that T_{ii} is an invariant.

2. $\bar{x} = Ax$, $\bar{\bar{x}} = B\bar{x}$ are two successive orthogonal transformations relative to each of which T_{ij} transforms as a tensor. Show that the resultant transformation $\bar{\bar{x}} = BAx$ is orthogonal and that T_{ij} transforms as a tensor with respect to it.

3. If A_i, B_i are vectors and $X_{ij}A_iB_j$ is an invariant, prove that X_{ij} is a tensor.

4. Verify that the transformation

$$\bar{x}_1 = \frac{1}{15}(5x_1 - 14x_2 + 2x_3)$$

$$\bar{x}_2 = -\frac{1}{3}(2x_1 + x_2 + 2x_3)$$

$$\bar{x}_3 = \frac{1}{15}(10x_1 + 2x_2 - 11x_3)$$

is orthogonal. A vector field is defined in the x-frame by the equations $A_1 = x_1^2$, $A_2 = x_2^2$, $A_3 = x_3^2$. Calculate the field in the $\bar{x}$-frame and verify that $\operatorname{div} \mathbf{A}$ is an invariant.

5. A_{ijk} is a tensor, all of whose components are zero, except for the following: $A_{111} = A_{222} = 1, A_{212} = -2$. Calculate the components of the vector A_{iji}. Show that the transformation

$$\bar{x}_1 = \tfrac{1}{7}(-3x_1 - 6x_2 - 2x_3)$$
$$\bar{x}_2 = \tfrac{1}{7}(-2x_1 + 3x_2 - 6x_3)$$
$$\bar{x}_3 = \tfrac{1}{7}(6x_1 - 2x_2 - 3x_3)$$

is orthogonal and calculate the component $\bar{A}_{123}$ of the tensor in $\bar{x}$-frame. Write down the equations of the inverse transformation. If B_{ij} is a tensor whose components in the $\bar{x}$-frame all vanish except that $\bar{B}_{13} = 1$, calculate B_{12}. (Ans. $(-1, 1, 0)$; 120/343; 6/49.)

6. If $A = (I - B)(I + B)^{-1}$, where B is a skew-symmetric matrix, show that A is orthogonal. Taking

$$B = \begin{vmatrix} 0 & 2 & 2 \\ -2 & 0 & 0 \\ -2 & 0 & 0 \end{vmatrix}$$

calculate A and write down the rectangular Cartesian coordinate transformation equations $\bar{x} = Ax$. In the x-frame, the tensor C_{ij} is skew-symmetric and $C_{12} = C_{13} = 1, C_{23} = 0$. Calculate the component $\bar{C}_{12}$ in the $\bar{x}$-frame. In the $\bar{x}$-frame, all the components of the tensor $\bar{D}_{ijk}$ vanish except the following $\bar{D}_{121} = -1$, $\bar{D}_{122} = 2, \bar{D}_{123} = 5$. Calculate the component D_{111} in the x-frame. Calculate the components of the vectors D_{ijj} and $C_{ij}D_{ijk}$ in the x-frame. (Ans. $\bar{C}_{12} = 1; D_{111} = -980/729$; $(-14/9, -8/9, -8/9)$; $(35/9, -32/9, -7/9)$.)

7. A_{ij} is a tensor field defined in the x-frame by the equation $A_{ij} = x_i x_j$. Calculate its components at the point P where $x_1 = 0$, $x_2 = x_3 = 1$. The coordinates x_i of a point in the x-frame are related to the coordinates $\bar{x}_i$ of the same point in the $\bar{x}$-frame by the equations

$$\bar{x}_1 = \tfrac{1}{7}(-3x_1 - 6x_2 - 2x_3)$$
$$\bar{x}_2 = \tfrac{1}{7}(-2x_1 + 3x_2 - 6x_3)$$
$$\bar{x}_3 = \tfrac{1}{7}(6x_1 - 2x_2 - 3x_3)$$

Calculate the component $\bar{A}_{11}$ of the tensor field at P. In the x-frame, prove that $A_{ij,j} = 4x_i$, $A_{ij,ij} = 12$. (Ans. $\bar{A}_{11} = 64/49$.)

8. $\mathbf{x}$ is the position vector of a point P with respect to an origin O. OP is rotated through an angle θ about an axis whose direction is determined by the unit vector $\mathbf{u}$. If the new position vector of P is $\bar{\mathbf{x}}$, prove that

$$\bar{\mathbf{x}} = \mathbf{x}\cos\theta + (1 - \cos\theta)(\mathbf{x}\cdot\mathbf{u})\mathbf{u} + \mathbf{u}\times\mathbf{x}\sin\theta$$

Deduce that the coordinate transformation generated when a rectangular Cartesian frame $Ox_1x_2x_3$ is rotated through an acute angle $\sin^{-1}(4/5)$ about an axis through O having direction ratios $(1, 2, 2)$ to give a new frame $O\bar{x}_1\bar{x}_2\bar{x}_3$ is

$$45\bar{x}_1 = 29x_1 + 28x_2 - 20x_3$$
$$45\bar{x}_2 = -20x_1 + 35x_2 + 20x_3$$
$$45\bar{x}_3 = 28x_1 - 4x_2 + 35x_3$$

9. Verify that the transformation

$$\bar{x}_1 = \tfrac{1}{9}(x_1 - 8x_2 + 4x_3)$$
$$\bar{x}_2 = \tfrac{1}{9}(4x_1 + 4x_2 + 7x_3)$$
$$\bar{x}_3 = \tfrac{1}{9}(8x_1 - x_2 - 4x_3)$$

is orthogonal. In the x-frame, the tensor A_{ij} is skew-symmetric and $A_{12} = A_{13} = 1$, $A_{23} = 0$. Calculate the component $\bar{A}_{12}$ in the $\bar{x}$-frame. In the $\bar{x}$-frame, all the components of the tensor B_{ijk} vanish except the following: $\bar{B}_{121} = -1$, $\bar{B}_{122} = 2$, $\bar{B}_{123} = 5$. Calculate the component B_{111} in the x-frame. Calculate the components of the vectors B_{ijj} and $A_{ij}B_{ijk}$ in the x-frame. (Ans. $\bar{A}_{12} = 1/3$; $B_{111} = 188/729$; $2/9$, $-16/9$, $8/9$; $47/27$, $11/27$, $-10/27$.)

10. In the x-frame in $\mathscr{E}_3$ a tensor field is defined by the equation $A_{ijk} = x_i^2 + 2x_j^2 + x_k^2$. Calculate the divergence of the vector field A_{iji}. Also, calculate the curl of the vector field A_{ijj}. (Ans. $16(x_1 + x_2 + x_3)$; $6(x_2 - x_3)$, etc.)

11. A pair of rectangular Cartesian frames are related by the equations

$$\bar{x}_1 = \tfrac{1}{15}(5x_1 - 14x_2 + 2x_3)$$
$$\bar{x}_2 = -\tfrac{1}{3}(2x_1 + x_2 + 2x_3)$$
$$\bar{x}_3 = \tfrac{1}{15}(10x_1 + 2x_2 - 11x_3)$$

A_{ijk} is a tensor, all of whose components vanish in the x-frame except the following: $A_{111} = A_{222} = 2$, $A_{122} = 4$, $A_{233} = 13$. Calculate (a) the components of the vector A_{ijj} in the $\bar{x}$-frame, and (b) $\bar{A}_{123}$. If B_{ij} is a tensor whose components in the $\bar{x}$-frame all vanish except $\bar{B}_{12}$, $\bar{B}_{23}$, which are both unity, calculate B_{11}. If V is an invariant field given in the x-frame by $V = x_1^2$, calculate the field in the $\bar{x}$-frame and the components of grad V in this frame, where $\bar{x}_1 = \bar{x}_2 = \bar{x}_3 = 9$. (Ans. (a) $(-12, -9, 6)$; (b) $-1396/225$; $B_{11} = -2/3$; $\nabla V = (2, -4, 4)$.)

12. In the frame $O\,x_1 x_2 x_3$, all components of A_{ij} are zero except $A_{12} = -A_{21} = 1$. If the transformation equations to the $\bar{x}$-frame are

$$\bar{x}_1 = x_1 \cos\alpha + x_2 \sin\alpha$$
$$\bar{x}_2 = -x_1 \sin\alpha + x_2 \cos\alpha$$
$$\bar{x}_3 = x_3$$

prove that $\bar{A}_{12} = 1$.

13. If $A_{ij} = x_i^2 + x_j^2$ $(i, j = 1, 2, 3)$, prove that (a) $A_{ij,j} = 2(x_1 + x_2 + x_3 + x_i)$ (b) $A_{ij,ij} = 12$.

14. Verify that the coordinate transformation

$$25\bar{x}_1 = 9x_1 + 20x_2 + 12x_3$$
$$25\bar{x}_2 = 12x_1 - 15x_2 + 16x_3$$
$$5\bar{x}_3 = -4x_1 + 3x_3$$

is orthogonal and calculate the component $\bar{A}_{311}$ of the tensor A_{ijk} in the $\bar{x}$-frame if all its components in the x-frame vanish except for the following: $A_{123} = 25$, $A_{222} = -6$. All components of the tensor B_{ij} in the $\bar{x}$-frame are zero, except for the components $\bar{B}_{i3}$ $(i = 1, 2, 3)$. Show that, in the x-frame, all the components B_{i2} $(i = 1, 2, 3)$ vanish. (Ans. $-192/25$.)

15. (i) x is a column matrix such that $x'x = 1$. If $A = I - 2xx'$, where I is a unit matrix, prove that A is orthogonal. If $x' = \alpha(1, -2, 3)$ where α is a scalar multiplier, calculate α and hence find A. Written in matrix form, the coordinate transformation between two rectangular Cartesian frames is $\bar{x} = Ax$ (A is the matrix just calculated). In the x-frame, all the components of the tensor B_{ijkl} are zero, except that $B_{1133} = -20$, $B_{1232} = 29$. Calculate the component $\bar{B}_{2221}$ in the $\bar{x}$-frame. What is the value of $\bar{B}_{iijj}$? If $\mathfrak{C}_i$ is a pseudotensor with components $(2, 3, 6)$ in the x-frame, find its components in the $\bar{x}$-frame. (Ans. $\bar{B}_{2221} = 72/49$; $\bar{B}_{iijj} = -20$; $(0, -7, 0)$.)

(ii) $A_{ij}(x_1, x_2, x_3)$ is a tensor field. If $A_{ij} = \delta_{ij}x_2^2$, where δ_{ij} is the Kronecker delta, list the non-zero components of $\partial A_{ij}/\partial x_k$. Deduce the value of $\partial \bar{A}_{11}/\partial \bar{x}_1$ at the point where $\bar{x}_1 = 1$, $\bar{x}_2 = 2$, $\bar{x}_3 = 1$. (Take the frames to be related as in (i).) (Ans. $\partial \bar{A}_{11}/\partial \bar{x}_1 = 4/7$.)

16. Verify that the transformation

$$7\bar{x}_1 = 3x_1 + 6x_2 - 2x_3$$
$$7\bar{x}_2 = 2x_1 - 3x_2 - 6x_3$$
$$7\bar{x}_3 = 6x_1 - 2x_2 + 3x_3$$

is orthogonal and calculate the component $\bar{A}_{32}$ of the tensor A_{ij} in the $\bar{x}$-frame if all its components in the x-frame are zero except for the following: $A_{11} = 10$, $A_{22} = 29$. If the only non-zero component of the tensor B_{ijk} in the $\bar{x}$-frame is $\bar{B}_{232} = 343$, calculate the component B_{111} in the x-frame. If $\mathfrak{C}_{ij}$ is a pseudotensor and $\mathfrak{C}_{33} = 98$, all other components being zero, calculate $\bar{\mathfrak{C}}_{12}$. (Ans. $\bar{A}_{32} = 6$; $B_{111} = 24$; $\bar{\mathfrak{C}}_{12} = -24$.)

17. If A is a square skew-symmetric matrix, show that A^2 is symmetric. If, also, $A^3 = -A$, show that the matrix B given by $B = I + 2A^2$ is orthogonal. If

$$A = \begin{pmatrix} 0 & a & b \\ -a & 0 & c \\ -b & -c & 0 \end{pmatrix}$$

prove that $A^3 = -A$ provided $a^2 + b^2 + c^2 = 1$. Taking $a = 1/3$, $b = c = 2/3$, calculate the orthogonal matrix B. Written in matrix form, the coordinate transformation between two rectangular cartesian frames is $\bar{x} = Bx$. In the x-frame, the tensor C_{ij} has all its components equal to 1. Calculate the component $\bar{C}_{23}$ in the $\bar{x}$-frame. In the x-frame, a vector field A_i has components $A_1 = x_1^2 + x_2^2 + x_3^2$, $A_2 = A_3 = 0$. Obtain formulae for the field's components in the $\bar{x}$-frame. Calculate the divergence of the field in both frames and show that the results are equal. (Ans. $\bar{B}_{23} = 91/81$.)

18. (i) The equations

$$\bar{x}_1 = \tfrac{1}{9}(ax_1 + 8x_2 + 4x_3)$$
$$\bar{x}_2 = \tfrac{1}{9}(4x_1 + bx_2 + 7x_3)$$
$$\bar{x}_3 = \tfrac{1}{9}(8x_1 + x_2 + cx_3)$$

represent a transformation between rectangular Cartesian axes. Calculate the values of a, b, c. In the x-frame, all the components of the tensor A_{ij} are zero except $A_{23} = 9$. Calculate the component $\bar{A}_{31}$ in the $\bar{x}$-frame. (Ans. $a = 1$, $b = -4$, $c = -4$; $\bar{A}_{31} = 4/9$.)

(ii) If A_{ijkl} is a tensor, prove that A_{ijki} is also a tensor. In the x-frame referred to in (i), all the components of the tensor A_{ijkl} vanish except for the following: $A_{1131} = 4$, $A_{3133} = 5$, $A_{2222} = 18$. Calculate all the components of the tensor A_{ijki} in the x-frame and the $(1, 1)$-component in the $\bar{x}$-frame. (Ans. $44/3$).

19. Verify that the transformation

$$\bar{x}_1 = \tfrac{1}{25}(9x_1 + 20x_2 + 12x_3)$$
$$\bar{x}_2 = \tfrac{1}{25}(12x_1 - 15x_2 + 16x_3)$$
$$\bar{x}_3 = \tfrac{1}{5}(4x_1 - 3x_3)$$

is orthogonal and write down the inverse transformation. In the x-frame, all the components of the tensor A_{ijk} vanish except $A_{111} = 1$. Calculate the component $\bar{A}_{123}$ in the $\bar{x}$-frame. In the $\bar{x}$-frame, all the components of the tensor B_{ijk} vanish except that $\bar{B}_{212} = 1$, $\bar{B}_{313} = \bar{B}_{222} = \bar{B}_{121} = 2$. Calculate the components of the vector B_{iji} in the x-frame. (Ans. $\bar{A}_{123} = 432/3125$; $(3, 0, 4)$.)

20. Show that, for all angles α, β, the transformation

$$\bar{x}_1 = x_1 \cos \alpha \cos \beta + x_2 \cos \alpha \sin \beta - x_3 \sin \alpha$$
$$\bar{x}_2 = -x_1 \sin \beta + x_2 \cos \beta$$
$$\bar{x}_3 = x_1 \sin \alpha \cos \beta + x_2 \sin \alpha \sin \beta + x_3 \cos \alpha$$

is orthogonal. Obtain the form taken by the transformation when $\alpha = \beta = \pi/4$ and, in this case, calculate the component $\bar{A}_{1123}$ of the tensor A_{ijkl} in the $\bar{x}$-frame, if the only non-zero components in the x-frame are $A_{1123} = A_{2213} = 1$. (Ans. 0.)

21. If A, B are orthogonal matrices of the same order, prove that AB is orthogonal. If

$$A = \tfrac{1}{5} \begin{pmatrix} 3 & 4 & 0 \\ -4 & 3 & 0 \\ 0 & 0 & 5 \end{pmatrix}, \qquad B = \tfrac{1}{5} \begin{pmatrix} 3 & 0 & 4 \\ 0 & 5 & 0 \\ -4 & 0 & 3 \end{pmatrix}$$

calculate the orthogonal matrix $C = AB$. The coordinates x_i and $\bar{x}_i$ in two rectangular Cartesian frames are related by the transformation $\bar{x} = Cx$. All the components of the tensor A_{ijk} vanish in the x-frame except $A_{221} = 1$, $A_{122} = 25$. Calculate the component $\bar{A}_{321}$ in the $\bar{x}$-frame. In the $\bar{x}$-frame, the only non-zero components of the tensor B_{ijkl} are $\bar{B}_{3211} = -5$, $\bar{B}_{3212} = 10$, $\bar{B}_{3213} = 15$. Calculate the components of the vector $A_{ijk}B_{ijkl}$ in the $\bar{x}$-frame and deduce its components in the x-frame. (Ans. $\bar{A}_{321} = -48/5$; $4464/25$, $-96/5$, $-48/25$.)

22. If $\mathfrak{A}_{ij}$ is a pseudotensor, $\mathfrak{B}_i$ is a pseudovector and E_i is a vector, prove that $\mathfrak{A}_{ij}\mathfrak{B}_i E_j$ is an invariant.

23. A vector field has components in the x-frame given by

$$A_1 = \sin(x_2 x_3), \ A_2 = \cos(x_3 x_1), \ A_3 = \tan(x_1 x_2)$$

Calculate curl $\mathbf{A}$ at the point $x_1 = x_2 = x_3 = \tfrac{1}{2}\sqrt{\pi}$. If the $\bar{x}$-frame is obtained from the x-frame by reversing the sense of the x_2-axis, what are the components of curl$\mathbf{A}$ in the $\bar{x}$-frame? (Ans. $\sqrt{\pi}(1 + 1/2\sqrt{2})$, $-\sqrt{\pi}(1 - 1/2\sqrt{2})$, $-\tfrac{1}{2}\sqrt{(2\pi)}$; $-\sqrt{\pi}(1 + 1/2\sqrt{2})$, $-\sqrt{\pi}(1 - 1/2\sqrt{2})$, $\tfrac{1}{2}\sqrt{(2\pi)}$.)

24. x_i, $\bar{x}_i$ $(i = 1, 2, 3)$ are coordinates of the same point with respect to two different rectangular Cartesian frames. If

$$\bar{x}_1 = \alpha(ax_1 + 2x_2 + 5x_3)$$
$$\bar{x}_2 = \beta(x_1 + bx_2 + 2x_3)$$
$$\bar{x}_3 = \gamma(2x_1 - 11x_2 + cx_3)$$

where α, β, γ are all positive, calculate the values of α, β, γ, a, b, c. If A_{ijkl} is a tensor whose only non-zero components in the x-frame are $A_{1313} = 1$, $A_{3223} = 20$, calculate its component $\bar{A}_{3121}$ in the $\bar{x}$-frame. In the $\bar{x}$-frame, the only non-zero components of the tensors A_i, B_{ij} are as follows: $\bar{A}_1 = 3$, $\bar{A}_2 = 1$, $\bar{B}_{13} = 5$, $\bar{B}_{22} = 3$. Calculate the components of the tensor $A_i B_{ij}$ in the x-frame. (Ans. $a = -14$, $b = 2$, $c = 10$, $\alpha = 1/15$, $\beta = 1/3$, $\gamma = 1/15$; $\bar{A}_{3121} = 2/5$; $(3, -9, 12)$.)

25. A is an anti-symmetric matrix such that $A^3 = -A$. Prove that the matrix B

$= I + A + A^2$ is orthogonal. Show that the matrix

$$A = \tfrac{1}{3} \begin{pmatrix} 0 & 1 & 2 \\ -1 & 0 & 2 \\ -2 & -2 & 0 \end{pmatrix}$$

satisfies the stated conditions and hence calculate the orthogonal matrix B. The coordinates of a point relative to a pair of rectangular Cartesian frames are related by the matrix transformation $\bar{x} = Bx$. A tensor A_{ij} has all its components zero in the x-frame except that $A_{32} = 81$. Calculate the component $\bar{A}_{12}$ in the $\bar{x}$-frame. If the tensor B_{ijk} has all its components zero in the $\bar{x}$-frame except $\bar{B}_{123} = 729$, calculate the component B_{321} in the x-frame. (Ans. $\bar{A}_{12} = 32$; $B_{321} = -128$.)

26. In $\mathscr{E}_3$, prove that

$$\text{curl grad } V = 0, \quad \text{div curl } \mathbf{A} = 0.$$

27. In $\mathscr{E}_3$, prove that

$$\text{(i) } e_{ikl}\, e_{imn} = \delta_{km}\delta_{ln} - \delta_{kn}\delta_{lm}$$

$$\text{(ii) } e_{ikl}\, e_{ikm} = 2\delta_{lm}$$

28. In $\mathscr{E}_3$, show that

$$\nabla^2 V = \text{div grad } V = \frac{\partial^2 V}{\partial x_1^2} + \frac{\partial^2 V}{\partial x_2^2} + \frac{\partial^2 V}{\partial x_3^2} = \frac{\partial^2 V}{\partial x_i \partial x_i}$$

29. In $\mathscr{E}_3$, prove that

$$\text{curl curl } \mathbf{A} = \text{grad div } \mathbf{A} - \nabla^2 \mathbf{A}$$

(*Hint:* Employ Exercise 27(i).)

30. In $\mathscr{E}_3$, prove that

(i) $$\mathbf{A} \times (\mathbf{B} \times \mathbf{C}) = \mathbf{A} \cdot \mathbf{C}\,\mathbf{B} - \mathbf{A} \cdot \mathbf{B}\,\mathbf{C}$$

(ii) $$\mathbf{A} \cdot \mathbf{B} \times \mathbf{C} = \begin{vmatrix} A_1 & B_1 & C_1 \\ A_2 & B_2 & C_2 \\ A_3 & B_3 & C_3 \end{vmatrix}$$

31. In $\mathscr{E}_N$, prove that

$$\text{div } V\mathbf{A} = V \text{ div } \mathbf{A} + \mathbf{A} \cdot \text{grad } V$$

32. In $\mathscr{E}_3$, prove that

(i) $\text{curl } V\mathbf{A} = V \text{ curl } \mathbf{A} - \mathbf{A} \times \text{grad } V$

(ii) $\text{div}(\mathbf{A} \times \mathbf{B}) = \mathbf{B} \cdot \text{curl } \mathbf{A} - \mathbf{A} \cdot \text{curl } \mathbf{B}$

(iii) $\text{curl}(\mathbf{A} \times \mathbf{B}) = \mathbf{B} \cdot \nabla \mathbf{A} - \mathbf{A} \cdot \nabla \mathbf{B} + \mathbf{A} \text{ div } \mathbf{B} - \mathbf{B} \text{ div } \mathbf{A}$

(iv) $\text{grad}(\mathbf{A} \cdot \mathbf{B}) = \mathbf{B} \cdot \nabla \mathbf{A} + \mathbf{A} \cdot \nabla \mathbf{B} + \mathbf{A} \times \text{curl } \mathbf{B} + \mathbf{B} \times \text{curl } \mathbf{A}$

where $\qquad \mathbf{A} \cdot \nabla \mathbf{B} = A_j B_{i,\,j}.$

33. If A_{ij} is a tensor and $B_{ij} = A_{ji}$, prove that B_{ij} is a tensor. Deduce that if A_{ij} is symmetric in one frame, it is so in all.

34. Prove that $\qquad \delta_{ij}\delta_{ik} = \delta_{jk}$

and that $\mathbf{e}_{ijk}\mathbf{e}_{lmn}$ has the value $+1$ if i, j, k are all different and (lmn) is an even permutation of (ijk), -1 if i, j, k are all different and (lmn) is an odd permutation of (ijk), and 0 otherwise. Deduce that

$$\mathbf{e}_{ijk}\mathbf{e}_{lmn} = \delta_{il}\delta_{jm}\delta_{kn} + \delta_{im}\delta_{jn}\delta_{kl} + \delta_{in}\delta_{jl}\delta_{km}$$
$$- \delta_{in}\delta_{jm}\delta_{kl} - \delta_{il}\delta_{jn}\delta_{km} - \delta_{im}\delta_{jl}\delta_{kn}$$

Hence prove that

$$\mathbf{e}_{ijk}\mathbf{e}_{imn} = \delta_{jm}\delta_{kn} - \delta_{jn}\delta_{km}$$

35. In $\mathscr{E}_3$, prove that

(i) $\qquad (\mathbf{a} \times \mathbf{b}) \cdot (\mathbf{c} \times \mathbf{d}) = \mathbf{a} \cdot \mathbf{c}\,\mathbf{b} \cdot \mathbf{d} - \mathbf{a} \cdot \mathbf{d}\,\mathbf{b} \cdot \mathbf{c}$

(ii) $\qquad (\mathbf{a} \times \mathbf{b}) \times (\mathbf{c} \times \mathbf{d}) = [\mathbf{acd}]\mathbf{b} - [\mathbf{bcd}]\mathbf{a}$
$$= [\mathbf{abd}]\mathbf{c} - [\mathbf{abc}]\mathbf{d}$$

where $[\mathbf{abc}] = \mathbf{a} \cdot \mathbf{b} \times \mathbf{c}$.

CHAPTER 3

Special Relativity Mechanics

15. The velocity vector

Suppose that a point P is in motion relative to an inertial frame S. Let ds be the distance between successive positions of P which it occupies at times t, $t + dt$ respectively. Then, by equation (7.4), if $d\tau$ is the proper time interval between these two events,

$$d\tau = \left(dt^2 - \frac{1}{c^2} ds^2 \right)^{1/2} = (1 - v^2/c^2)^{1/2} dt \qquad (15.1)$$

where $v = ds/dt$ is the speed of P as measured in S. Now, as shown in section 7, $d\tau$ is the time interval between the two events as measured in a frame for which the events occur at the same point. Thus $d\tau$ is the time interval measured by a clock moving with P. dt is the time interval measured by clocks stationary in S. Equation (15.1) indicates that, as observed from S, the rate of the clock moving with P is slow by a factor $(1 - v^2/c^2)^{1/2}$. This is the phenomenon of time dilation already commented upon in section 6. If P leaves a point A at $t = t_1$ and arrives at a point B at $t = t_2$, the time of transit as registered by a clock moving with P will be

$$\tau_2 - \tau_1 = \int_{t_1}^{t_2} (1 - v^2/c^2)^{1/2} dt \qquad (15.2)$$

The successive positions of P together with the times it occupies these positions constitute a series of events which will lie on the point's world-line in Minkowski space–time. Erecting rectangular axes in space–time corresponding to the rectangular Cartesian frame S, let x_i, $x_i + dx_i$ be the coordinates of adjacent points on the world-line. These points will represent the events (x, y, z, t), $(x + dx, y + dy, z + dz, t + dt)$ in S. If (v_x, v_y, v_z) are the components of the velocity vector $\mathbf{v}$ of P relative to S, then

$$v_x = \frac{dx}{dt}, \quad v_y = \frac{dy}{dt}, \quad v_z = \frac{dz}{dt} \qquad (15.3)$$

$\mathbf{v}$ does not possess the transformation properties of a vector relative to orthogonal transformations (i.e. Lorentz transformations) in space–time. It is a vector relative to rectangular axes stationary in S only. However, we can define a 4-*velocity vector* which does possess such properties as follows: dx_i is a displacement vector relative to rectangular axes in space–time and $d\tau$ is an invariant. It follows that $dx_i/d\tau$ is a vector relative to Lorentz transformations expressed as orthogonal transformations in space–time. It is called the 4-velocity vector of P and will be denoted by **V**.

V can be expressed in terms of **v** thus:

$$\frac{dx_i}{d\tau} = \frac{dx_i}{dt}\frac{dt}{d\tau} = (1 - v^2/c^2)^{-1/2}\dot{x}_i \tag{15.4}$$

by equation (15.1). Also, from equations (4.4) we obtain

$$\dot{x}_1 = v_x, \quad \dot{x}_2 = v_y, \quad \dot{x}_3 = v_z, \quad \dot{x}_4 = ic \tag{15.5}$$

It now follows from these equations that

$$\mathbf{V} = (1 - v^2/c^2)^{-1/2}(v_x, v_y, v_z, ic) = (1 - v^2/c^2)^{-1/2}(\mathbf{v}, ic) \tag{15.6}$$

where the notation should be clear without further explanation.

Knowing the manner in which the components of **V** transform when new axes are chosen in space–time, equation (15.6) enables us to calculate how the components of **v** transform when S is replaced by a new inertial frame $\bar{S}$. Thus, consider the orthogonal transformation (5.1) which has been interpreted as a change from an inertial frame S to another $\bar{S}$ related to the first as shown in Fig. 2. The corresponding transformation equations for **V** are

$$\left.\begin{array}{ll} \bar{V}_1 = V_1 \cos\alpha + V_4 \sin\alpha & \bar{V}_2 = V_2 \\ \bar{V}_4 = -V_1 \sin\alpha + V_4 \cos\alpha & \bar{V}_3 = V_3 \end{array}\right\} \tag{15.7}$$

By equation (15.6), these equations are equivalent to

$$\left.\begin{array}{l} (1 - \bar{v}^2/c^2)^{-1/2}\bar{v}_x = (1 - v^2/c^2)^{-1/2}(v_x \cos\alpha + ic \sin\alpha) \\ (1 - \bar{v}^2/c^2)^{-1/2}\bar{v}_y = (1 - v^2/c^2)^{-1/2}v_y \\ \\ (1 - \bar{v}^2/c^2)^{-1/2}\bar{v}_z = (1 - v^2/c^2)^{-1/2}v_z \\ (1 - \bar{v}^2/c^2)^{-1/2}ic = (1 - v^2/c^2)^{-1/2}(-v_x \sin\alpha + ic \cos\alpha) \end{array}\right\} \tag{15.8}$$

where $\bar{v}$ is the velocity of the point as measured in the frame $\bar{S}$. Substituting for $\cos\alpha$, $\sin\alpha$ from equations (5.7), equations (15.8) can be written

$$\left.\begin{array}{l} \bar{v}_x = Q(v_x - u) \\ \bar{v}_y = Q(1 - u^2/c^2)^{1/2}v_y \\ \\ \bar{v}_z = Q(1 - u^2/c^2)^{1/2}v_z \\ 1 = Q(1 - uv_x/c^2) \end{array}\right\} \tag{15.9}$$

where

$$Q = \left[\frac{1 - \bar{v}^2/c^2}{(1 - v^2/c^2)(1 - u^2/c^2)}\right]^{1/2} \tag{15.10}$$

Dividing the first three equations (15.9) by the fourth, we obtain the special Lorentz transformation equations for **v** in their final form, viz.

$$\left.\begin{aligned}
\bar{v}_x &= \frac{v_x - u}{1 - uv_x/c^2} \\
\bar{v}_y &= \frac{(1 - u^2/c^2)^{1/2}v_y}{1 - uv_x/c^2} \\
\bar{v}_z &= \frac{(1 - u^2/c^2)^{1/2}v_z}{1 - uv_x/c^2}
\end{aligned}\right\} \tag{15.11}$$

If u and v are small by comparison with c, equations (15.11) can be replaced by the approximate equations

$$\bar{v}_x = v_x - u, \quad \bar{v}_y = v_y, \quad \bar{v}_z = v_z \tag{15.12}$$

These are equivalent to the vector equation (1.1) relating velocity measurements in two inertial frames according to Newtonian mechanics.

Since, by the fourth of equations (15.9), Q must be real, equation (15.10) implies that if $\bar{v} < c$ then $v < c$. Thus, if a point is moving with a velocity approaching c in $\bar{S}$ and $\bar{S}$ is moving relative to S with a velocity of the same order, the point's velocity relative to S will still be less than c. Such a result is, of course, completely at variance with classical ideas. In particular, if a light pulse is being propagated along Ox so that $v_x = c, v_y = v_z = 0$, then it will be found that $\bar{v}_x = c, \bar{v}_y = \bar{v}_z = 0$. This confirms that light is propagated with speed c in all inertial frames.

The transformation inverse to (15.11) can be found by exchanging 'barred' and 'unbarred' velocity components and replacing u by $-u$.

Suppose that particles A and B move along the x-axis of a frame S with speeds $3c/4$ in opposite directions, both leaving O at time $t = 0$. At any later time t, their x-coordinates will be $x_A = -3ct/4, x_B = 3ct/4$ and their distance apart will be $x_B - x_A = 3ct/2$. Clearly, this distance increases at a rate $3c/2$. However, this is not their relative velocity and the fact that the rate exceeds c does not conflict with the result already derived that no material body can be observed from any inertial frame to have a speed greater than c. There is no special-relativity prohibition against the distance between two bodies increasing at a rate greater than c. To find the velocity of B relative to A, it is necessary to introduce a second inertial frame $\bar{S}$ with its origin at A; the velocity of B in $\bar{S}$ is then the velocity of B relative to A. Since the velocity of $\bar{S}$ relative to S is $u = -3c/4$ and the velocity of B in S is $v_x = 3c/4$, the first of equations (15.11) gives $\bar{v}_x = 24c/25$ ($< c$) as the velocity of B relative to A.

16. Mass and momentum

In section 2 it was shown that Newton's laws of motion conform to the special principle of relativity. However, the argument involved classical ideas concerning space–time relationships between two inertial frames and these have since been replaced by relationships based upon the Lorentz transformation. The whole question must therefore be re-examined and this we shall do in this and the following section.

We shall begin by considering the conservation of momentum, equation (1.3), for the impact of two particles by which mass is defined in classical mechanics. Since the velocity vectors $\mathbf{u}_1$, etc. are not vectors relative to orthogonal transformations in space–time, and indeed transform between inertial frames in a very complex manner, it is at once evident that equation (1.3) is not covariant with respect to transformations between inertial frames. It will accordingly be replaced, tentatively, by another equation, viz.

$$M_1 \mathbf{U}_1 + M_2 \mathbf{U}_2 = M_1 \mathbf{V}_1 + M_2 \mathbf{V}_2 \tag{16.1}$$

where $\mathbf{U}_1$, etc. are the 4-velocities of the particles and M_1, M_2 are invariants associated with the particles which will correspond to their classical masses. This is a vector equation and hence is covariant with respect to orthogonal transformations in space–time as we require. Equation (16.1) will be abbreviated to the statement

$$\Sigma\, M\, \mathbf{V} \text{ is conserved} \tag{16.2}$$

and then, by equation (15.6), this implies that

$$\Sigma\, m(\mathbf{v}, ic) \text{ is conserved} \tag{16.3}$$

where
$$m = \frac{M}{(1 - v^2/c^2)^{1/2}} \tag{16.4}$$

By consideration of the first three (or space) components of (16.3), it will be clear that

$$\Sigma\, m\mathbf{v} \text{ is conserved} \tag{16.5}$$

and, by consideration of the fourth (or time) component that

$$\Sigma\, m \text{ is conserved} \tag{16.6}$$

If, therefore, m is identified as the quantity which will play the role of the Newtonian mass in special relativity mechanics, our tentative conservation law (16.1) is seen to incorporate both the principles of conservation of momentum and of mass from Newtonian mechanics. The principle (16.1) is accordingly eminently reasonable. However, our ultimate justification for accepting it is, of course, that its consequences are verified experimentally. We shall refer to such checks at appropriate points in the later development.

It appears from equation (16.4) that the mass of a particle must now be regarded as being dependent upon its speed v. If $v = 0$, then $m = M$. Thus M is the mass of the particle when measured in an inertial frame in which it is stationary. M will be referred to as the *rest mass* or *proper mass* and will, in future, be denoted by m_0. To distinguish it from m_0, m is often called the *inertial mass*. Then

$$m = \frac{m_0}{(1 - v^2/c^2)^{1/2}} \qquad (16.7)$$

Clearly $m \to \infty$ as $v \to c$, implying that inertia effects become increasingly serious as the velocity of light is approached and prevent this velocity being attained by any material particle. This is in agreement with our earlier observations. Formula (16.7) has been verified by observation of collisions between atomic nuclei and cosmic ray particles (e.g., see Exercise 27 at the end of this chapter).

We shall define the 4-*momentum vector* **P** of a particle whose rest mass is m_0 and whose 4-velocity is **V**, by the equation

$$\mathbf{P} = m_0\,\mathbf{V} \qquad (16.8)$$

Since m_0 is an invariant and **V** is a vector in space–time, **P** is a vector. By equation (15.6),

$$\mathbf{P} = m_0\,(1 - v^2/c^2)^{-1/2}\,(\mathbf{v}, ic) = (m\mathbf{v}, imc) = (\mathbf{p}, imc) \qquad (16.9)$$

where $\mathbf{p} = m\mathbf{v}$ is the classical momentum.

Relative to the special orthogonal transformation (5.1), the transformation equations for the components of **P** are

$$\left. \begin{aligned} \bar{P}_1 &= P_1 \cos\alpha + P_4 \sin\alpha \qquad \bar{P}_2 = P_2 \\ \bar{P}_4 &= -P_1 \sin\alpha + P_4 \cos\alpha \qquad \bar{P}_3 = P_3 \end{aligned} \right\} \qquad (16.10)$$

Substituting for the components of **P** from equation (16.9) and similarly for **P̄**, and employing equations (5.7), it will be found that

$$\left. \begin{aligned} \bar{p}_x &= \frac{p_x - mu}{(1 - u^2/c^2)^{1/2}} \\ \bar{p}_y &= p_y \\ \bar{p}_z &= p_z \end{aligned} \right\} \qquad (16.11)$$

$$\bar{m} = \frac{m - p_x u/c^2}{(1 - u^2/c^2)^{1/2}} \qquad (16.12)$$

Equations (16.11) constitute the special Lorentz transformation equations for the components of the momentum **p** and equation (16.12) the corresponding transformation equation for mass. Since $p_x = mv_x$, this equation can also be written

$$\bar{m} = \frac{1 - uv_x/c^2}{(1 - u^2/c^2)^{1/2}}\,m \qquad (16.13)$$

This reduces to the classical form of equation (2.4) if u, v_x are negligible by comparison with c.

17. The force vector. Energy

We have seen that in classical mechanics, when the mass of a particle has been determined, the force acting upon it at any instant is specified by Newton's second law. Force receives a similar defintion in special relativity mechanics. The mass of a particle with a given velocity can be determined by permitting it to collide with a standard particle and applying the principle of momentum conservation. Equation (16.7) then gives its mass at any velocity. The force $\mathbf{f}$ acting upon a particle having mass m and velocity $\mathbf{v}$ relative to some inertial frame is then defined by the equation

$$\mathbf{f} = \frac{d}{dt}(m\mathbf{v}) = \frac{d\mathbf{p}}{dt} \tag{17.1}$$

where $\mathbf{p}$ is the particle's momentum. Clearly $\mathbf{f}$ will be dependent upon the inertial frame employed, a departure from classical mechanics.

Definition (17.1) implies that, if equal and opposite forces act upon two colliding particles, momentum is conserved. However, although experiment confirms that momentum is indeed conserved, Newton's third law cannot be incorporated in the new mechanics, for it will appear later that, if the forces are equal and opposite for one inertial observer, in general they are not so for all such observers. Equation (16.1) therefore replaces this law in the new mechanics.

$\mathbf{f}$ is not a vector with respect to Lorentz transformations in space–time. However, a 4-*force* $\mathbf{F}$ can be defined which has this property. The natural definition is clearly

$$\mathbf{F} = \frac{d\mathbf{P}}{d\tau} = m_0 \frac{d\mathbf{V}}{d\tau} \tag{17.2}$$

$\mathbf{P}$ being the 4-momentum and τ the proper time for the particle. $\mathbf{F}$ is immediately expressible in terms of $\mathbf{f}$ for, by equation (16.9),

$$\mathbf{F} = \frac{d}{d\tau}(\mathbf{p}, imc)$$

$$= \frac{d}{dt}(\mathbf{p}, imc)\frac{dt}{d\tau}$$

$$= (1 - v^2/c^2)^{-1/2}(\dot{\mathbf{p}}, i\dot{m}c)$$

$$= (1 - v^2/c^2)^{-1/2}(\mathbf{f}, i\dot{m}c) \tag{17.3}$$

The vectors $\mathbf{V}$, $\mathbf{F}$ are orthogonal. This is proved as follows: From equation (15.6),

$$\mathbf{V}^2 = -c^2 \tag{17.4}$$

Differentiating with respect to τ,

$$\mathbf{V} \cdot \frac{d\mathbf{V}}{d\tau} = 0$$

i.e.
$$\mathbf{V} \cdot \mathbf{F} = 0 \tag{17.5}$$

as stated. This result has very important consequences. Substituting for $\mathbf{V}$ and $\mathbf{F}$ from equations (15.6) and (17.3) respectively, it is clear that

$$(1 - v^2/c^2)^{-1} (\mathbf{v}, ic) \cdot (\mathbf{f}, i\dot{m}c) = 0. \tag{17.6}$$

This is equivalent to
$$\mathbf{v} \cdot \mathbf{f} - c^2 \dot{m} = 0. \tag{17.7}$$

But, by definition, $\mathbf{v} \cdot \mathbf{f}$ is the rate at which $\mathbf{f}$ is doing work. It follows that the work done by the force acting on the particle during a time interval (t_1, t_2) is

$$\int_{t_1}^{t_2} c^2 \dot{m} dt = m_2 c^2 - m_1 c^2 \tag{17.8}$$

The classical equation of work is

$$\text{work done} = \text{increase in kinetic energy} \tag{17.9}$$

where $T = \frac{1}{2}mv^2$ is the kinetic energy (KE). Equation (17.8) indicates that in special-relativity mechanics we must define T by a formula of the type

$$T = mc^2 + \text{constant} \tag{17.10}$$

When $v = 0, T = 0$ and this determines the unknown constant to be $-m_0 c^2$. Thus

$$T = \frac{m_0 c^2}{(1 - v^2/c^2)^{1/2}} - m_0 c^2 \tag{17.11}$$

If v/c is small $(1 - v^2/c^2)^{-1/2} = 1 + v^2/2c^2$ approximately and the above equation reduces to $T = \frac{1}{2}m_0 v^2$, in agreement with classical theory.

According to equation (17.10), any increase in the kinetic energy of a particle will result in a proportional increase in its mass. Thus, if a body is heated so that the thermal agitation of its molecules is increased, the masses of these particles, and hence the total body mass, will increase in proportion to the heat energy which has been communicated.

Again, suppose two equal elastic particles approach one another along the same straight line with equal speeds v. If their rest masses are both m_0, the net mass in the system before collision is

$$2m_0/(1 - v^2/c^2)^{1/2}$$

It has been accepted as a fundamental principle that this mass will be conserved during the collision. However, from considerations of symmetry, it is obvious that at some instant during the impact both particles will be brought to rest and their masses at this instant will be proper masses m_0'. By our principle,

$$2m_0' = \frac{2m_0}{(1 - v^2/c^2)^{1/2}} \tag{17.12}$$

It follows, therefore, that, at this instant, the rest mass of each particle has increased by

$$\frac{m_0}{(1 - v^2/c^2)^{1/2}} - m_0 = T/c^2 \tag{17.13}$$

where T is the original KE of the particle and use has been made of equation (17.11). Now, in losing this KE, the particle has had an equal amount of work done upon it by the force of interaction and this has resulted in a distortion of the elastic material of which it is made. At the instant each particle is brought to rest, this distortion is at a maximum and the elastic potential energy as measured by the work done will be exactly T. If we assume that this increase in the internal energy of the particle leads to a proportional increase in mass, the increment of rest mass (17.13) is explained. If the particles are not perfectly elastic, the work done in bringing them to rest will not only increase the internal elastic energy, but will also generate heat. Both forms of energy will then contribute to increase the rest masses.

Such considerations as these suggest very strongly that mass and energy are equivalent, being two different measures of the same physical quantity. Thus, the distinction between mass and energy which was maintained in classical physical theories, has now been abandoned. All forms of energy E, mechanical, thermal, electromagnetic, are now taken to possess inertia of mass m, according to *Einstein's equation*, viz.

$$E = mc^2 \tag{17.14}$$

Conversely, any particle whose mass is m, has associated energy E and, by equation (17.11),

$$E = T + m_0 c^2 \tag{17.15}$$

$m_0 c^2$ is interpreted as the internal energy of the particle when stationary. If the particle were converted completely into electromagnetic radiation, $m_0 c^2$ would be the energy released. This is the source of the energy released in an atomic explosion. The mass of the material products of the explosion is slightly less than the net mass present before the explosion, the difference being accounted for by the mass of the energy released. Even a small mass deficiency implies that an immense quantity of energy has been released. Thus, if $m = 1$ kg, $c = 3 \times 10^8$ m s^{-1}, then $E = 9 \times 10^{16}$ J $= 2.5 \times 10^{10}$ kWh.

The principle of conservation of mass, which has been incorporated into the new mechanics, is now seen to be identical with the principle of conservation of energy, which is accordingly also regarded as valid in the new mechanics. However, the distinction between the two principles, which was a feature of the older mechanics, has disappeared.

18. Lorentz transformation equations for force

By equation (17.7),

$$\dot{i m} c = \frac{i}{c} \mathbf{f} \cdot \mathbf{v} \tag{18.1}$$

Referring to equation (17.3), **F** can now be completely expressed in terms of **f**; thus

$$\mathbf{F} = (1 - v^2/c^2)^{-1/2} \left(\mathbf{f}, \frac{i}{c} \mathbf{f} \cdot \mathbf{v} \right) \tag{18.2}$$

Relative to the special Lorentz transformation, the transformation equations for the components of **F** are

$$\begin{aligned}
\overline{F}_1 &= F_1 \cos \alpha + F_4 \sin \alpha & \overline{F}_2 &= F_2 \\
\overline{F}_4 &= -F_1 \sin \alpha + F_4 \cos \alpha & \overline{F}_3 &= F_3
\end{aligned} \tag{18.3}$$

Substituting from equation (18.2) into the first three of these equations and employing equations (5.7), it follows that

$$\begin{aligned}
\overline{f}_x &= Q\left(f_x - \frac{u}{c^2} \mathbf{f} \cdot \mathbf{v} \right) \\
\overline{f}_y &= Q(1 - u^2/c^2)^{1/2} f_y \\
\overline{f}_z &= Q(1 - u^2/c^2)^{1/2} f_z
\end{aligned} \tag{18.4}$$

where Q is given by equation (15.10). Substituting for Q from the fourth of equations (15.9), it will be found that

$$\begin{aligned}
\overline{f}_x &= f_x - \frac{u}{c^2} \cdot \frac{(f_y v_y + f_z v_z)}{1 - uv_x/c^2} \\
\overline{f}_y &= \frac{(1 - u^2/c^2)^{1/2}}{1 - uv_x/c^2} f_y \\
\overline{f}_z &= \frac{(1 - u^2/c^2)^{1/2}}{1 - uv_x/c^2} f_z
\end{aligned} \tag{18.5}$$

These are the special Lorentz transformation equations for **f**. If u, v are negligible by comparison with c, these equations reduce to the classical form of equation (2.6).

It is clear from equations (18.5) that, if equal and opposite forces are observed from S to act upon two particles, the forces observed from $\overline{S}$ will not be so related unless the particles' velocities are the same.

19. Fundamental particles. Photon and neutrino

By eliminating m and v between equations (16.7), (17.14) and the equation $p = mv$ giving the linear momentum of a particle, it will be found that

$$E = c \sqrt{(p^2 + m_0^2 c^2)} \tag{19.1}$$

This useful equation relates the total energy E of a particle (including its internal energy) with its momentum p. A special case of great importance is when $m_0 = 0$, and then

$$E = cp \tag{19.2}$$

For such a particle, $m = E/c^2 = p/c$; i.e. its rest mass vanishes, but its inertial mass is non-zero. This result is inconsistent with equation (16.7), unless $v = c$ (in which case the right-hand member becomes indeterminate). We conclude that any particle having zero rest mass must always move with the speed of light.

Two such particles are known, the *photon* and the *neutrino*. The former is a quantum of electromagnetic energy and the latter is a particle which is generated in some interactions between fundamental particles governed by the *weak interaction force* (e.g. the decay of a neutron into a proton). Neither particle exhibits any electric charge. If either particle is absorbed by other matter, it loses its identity and delivers its energy and momentum to the absorbing body – the heating of a metal plate placed in sunlight is an example of the absorption of photons. Neutrinos are exceptionally difficult to absorb and hence to detect – there is a high probability that a neutrino from the sun will pass right through the earth without interaction with a single one of its atoms.

As an example of a particle interaction in which a neutrino is involved, consider the decay of a negative *pion* (a *meson*) into a *muon* (heavy electron) and a neutrino. Assuming that the pion is at rest in the laboratory frame, the momenta of the muon and neutrino will have equal magnitudes p, but will be in opposite senses. If m_π, m_μ are the proper masses of the pion and muon respectively, the principle of conservation of energy leads to the equation

$$m_\pi c^2 = cp + c \sqrt{(p^2 + m_\mu^2 c^2)} \qquad (19.3)$$

having used equations (19.1) and (19.2). Solving for p, we find

$$p = c(m_\pi^2 - m_\mu^2)/2m_\pi \qquad (19.4)$$

The energy of the muon is now found to be

$$E_\mu = m_\pi c^2 - cp = c^2(m_\pi^2 + m_\mu^2)/2m_\pi \qquad (19.5)$$

and the KE of this particle is accordingly

$$T_\mu = E_\mu - m_\mu c^2 = c^2(m_\pi - m_\mu)^2/2m_\pi \qquad (19.6)$$

From tables, we find that the rest masses in atomic energy units are $m_\pi c^2 = 140$ MeV, $m_\mu c^2 = 106$ MeV (mega-electron volts). Hence, $T_\mu = 4.1$ MeV. This value has been checked experimentally by observing the length of the path of the muon in the resistive medium in which it is generated (usually liquid hydrogen).

20. Lagrange's and Hamilton's equations

Suppose that a particle having constant rest mass m_0 is in motion relative to an inertial frame under the action of a force derivable from a potential V. Then its equations of motion are

$$\frac{d}{dt}\left\{ \frac{m_0 \dot{x}}{(1 - v^2/c^2)^{1/2}} \right\} = -\frac{\partial V}{\partial x}, \qquad \text{etc.} \qquad (20.1)$$

Expressed in Lagrange form, these equations must be

$$\frac{d}{dt}\left(\frac{\partial L}{\partial \dot{x}}\right) = \frac{\partial L}{\partial x}, \quad \text{etc.} \tag{20.2}$$

and hence L must be a function of $x, y, z, \dot{x}, \dot{y}, \dot{z}$, such that

$$\frac{\partial L}{\partial \dot{x}} = \frac{m_0 \dot{x}}{(1 - v^2/c^2)^{1/2}} \qquad \frac{\partial L}{\partial x} = -\frac{\partial V}{\partial x}, \quad \text{etc.} \tag{20.3}$$

Since $v^2 = \dot{x}^2 + \dot{y}^2 + \dot{z}^2$, these equations can be validated by taking

$$L = -m_0 c^2 (1 - v^2/c^2)^{1/2} - V \tag{20.4}$$

which is accordingly the Lagrangian for the particle.

Now

$$\frac{\partial L}{\partial \dot{x}} = p_x, \quad \text{etc.} \tag{20.5}$$

and it follows exactly as in classical theory that, if the Hamiltonian H is defined by the equation

$$H = p_x v_x + p_y v_y + p_z v_z - L \tag{20.6}$$

and is then expressed as a function of the quantities x, y, z, p_x, p_y, p_z alone, the Lagrange equations (20.2) are equivalent to Hamilton's equations

$$\dot{x} = \frac{\partial H}{\partial p_x}, \qquad \dot{p}_x = -\frac{\partial H}{\partial x}, \quad \text{etc.} \tag{20.7}$$

Now

$$p_x v_x + p_y v_y + p_z v_z = \frac{m_0 v^2}{(1 - v^2/c^2)^{1/2}} \tag{20.8}$$

and hence

$$\begin{aligned}
H &= \frac{m_0 v^2}{(1 - v^2/c^2)^{1/2}} + m_0 c^2 (1 - v^2/c^2)^{1/2} + V \\
&= \frac{m_0 c^2}{(1 - v^2/c^2)^{1/2}} + V \\
&= E + V
\end{aligned} \tag{20.9}$$

the total energy, precisely as for classical theory.

But

$$\begin{aligned}
p_x^2 + p_y^2 + p_z^2 &= \frac{m_0^2 v^2}{1 - v^2/c^2} \\
&= -m_0^2 c^2 + \frac{m_0^2 c^2}{1 - v^2/c^2} \\
&= -m_0^2 c^2 + E^2/c^2
\end{aligned} \tag{20.10}$$

and it follows that

$$E^2 = c^2(p_x^2 + p_y^2 + p_z^2 + m_0^2 c^2) \tag{20.11}$$

Substituting in equation (20.9),

$$H = c(p_x^2 + p_y^2 + p_z^2 + m_0^2 c^2)^{1/2} + V \tag{20.12}$$

expressing H as a function of x, y, z, p_x, p_y, p_z. The reader is now left to verify that Hamilton's equations are equivalent to the equations of motion (20.1).

21. Energy–momentum tensor

Suppose that there is a continuous distribution of mass over some region of space. In this section, we shall suppose this to take any physical form whatsoever. For example, the distribution may be in the form of the molecules of an elastic body and, in this case, the mass must include a component corresponding to the mass of the potential energy of the field of stress, in addition to the inertia of the particles themselves. Such a field will be electromagnetic in nature, the electric charges present in the molecules being ultimately responsible for its presence; we shall not, therefore, exclude a further contribution to the mass–energy distribution from any other electromagnetic field which may happen to be present. Any random motion of particles exhibiting itself as heat energy will also make a contribution. Equations governing this combined mass flow will now be derived.

Let S be an inertial frame $Ox_1 x_2 x_3$ and let μ', μ'', etc., be the densities of inertial mass at a point of the frame due to the various contributors. Then, if $\mathbf{v}'$, $\mathbf{v}''$, etc., are the respective velocities of flow of these components, the net density of linear momentum $\mathbf{g}$ will be given by

$$\mathbf{g} = \mu'\mathbf{v}' + \mu''\mathbf{v}'' + \ldots = \mu\mathbf{v}, \tag{21.1}$$

where

$$\mu = \mu' + \mu'' + \ldots \tag{21.2}$$

is the net density of inertial mass. Equation (21.1) defines the mean velocity of mass flow $\mathbf{v}$. In time dt, the mass flowing across an area dA having unit normal $\mathbf{n}$ is

$$\mu'\mathbf{v}' \cdot \mathbf{n} \, dA \, dt + \mu''\mathbf{v}'' \cdot \mathbf{n} \, dA \, dt + \ldots = \mu\mathbf{v} \cdot \mathbf{n} \, dA \, dt \tag{21.3}$$

thus, the rate of mass flow across unit area is $\mu\mathbf{v} \cdot \mathbf{n} = \mathbf{g} \cdot \mathbf{n}$, implying that $\mathbf{g}$ is also the current density vector for the mass flow. The components of $\mathbf{g}$ will be written g_α (Greek indices will range over values 1, 2, 3).

Let $\mathbf{g}^{(\alpha)}$ be the current density vector for the flow of the x_α-component of momentum, i.e. the rate of flow of this component of momentum across unit area with unit normal $\mathbf{n}$ is $\mathbf{g}^{(\alpha)} \cdot \mathbf{n}$. The x_β-component of $\mathbf{g}^{(\alpha)}$ will be denoted by $g_{\alpha\beta}$. In the special case of a cloud of non-interacting particles (no stress field), whose velocity of flow is $\mathbf{v}$, since the density of the x_α-component of momentum is g_α, the quantity of this component passing over the unit area in unit time is $g_\alpha \mathbf{v} \cdot \mathbf{n}$. It follows that

$\mathbf{g}^{(\alpha)} = g_\alpha \mathbf{v}$ and hence,

$$g_{\alpha\beta} = g_\alpha v_\beta \tag{21.4}$$

A simple distribution of this type will be referred to as an *incoherent cloud*.

For any distribution which includes material particles, in addition to the internal forces of interaction between the particles, there may be other forces acting upon them due to agents which are regarded as external to the system; such forces will be termed *external forces*. Let $d\omega$ be the volume occupied by a small element of the fluid or solid which is formed from these particles; if $\mathbf{v}$ is the flow velocity of the element and $d\omega_0$ is its proper volume (i.e. volume measured in a frame in which the element is momentarily stationary), then

$$d\omega = \sqrt{(1 - v^2/c^2)}\, d\omega_0 \tag{21.5}$$

since all lengths parallel to the flow will be subject to a Fitzgerald contraction. If $\mathbf{df}$ is the resultant external 3-force acting on the element, we shall define the 3-force density $\mathbf{d}$ at the element to be such that $\mathbf{df} = \mathbf{d}\, d\omega$. Similarly, if $\mathbf{dF}$ is the external 4-force on the element, the 4-force density will be $\mathbf{D}$, where $\mathbf{dF} = \mathbf{D}\, d\omega_0$; since $d\omega_0$ is a 4-invariant, this defines $\mathbf{D}$ as a 4-vector. Reference to equation (18.2) shows that we can write

$$\mathbf{dF} = (1 - v^2/c^2)^{-1/2}\, (\mathbf{df}, i\mathbf{df}\cdot\mathbf{v}/c) \tag{21.6}$$

and this is equivalent to the equation

$$\mathbf{D}\, d\omega_0 = (1 - v^2/c^2)^{-1/2}\, (\mathbf{d}, i\mathbf{d}\cdot\mathbf{v}/c)\, d\omega \tag{21.7}$$

It now follows from equation (21.5) that

$$\mathbf{D} = (\mathbf{d}, i\mathbf{d}\cdot\mathbf{v}/c) \tag{21.8}$$

which relates 3- and 4-force densities.

Now suppose Σ is a closed surface which is stationary relative to the frame S and let $d\sigma$ be the area of a surface element whose outwardly directed unit normal is $\mathbf{n}$. Then the rate of increase of the total mass inside Σ must equal the rate of inflow of mass across Σ, plus the rate at which external forces acting on any particles inside Σ generate energy (and hence mass) by performing work. Let Γ denote the interior of Σ and $d\omega$ a volume element of Γ. Then conservation of inertial mass is expressed by the equation.

$$\frac{d}{dt}\int_\Gamma \mu\, d\omega = -\int_\Sigma \mathbf{g}\cdot\mathbf{n}\, d\sigma + \frac{1}{c^2}\int_\Gamma \mathbf{d}\cdot\mathbf{v}\, d\omega \tag{21.9}$$

Converting the surface integral into a volume integral over Γ by the divergence theorem, this equation is seen to be equivalent to

$$\int_\Gamma \left\{ \frac{\partial\mu}{\partial t} + \operatorname{div}\mathbf{g} - \frac{1}{c^2}\mathbf{d}\cdot\mathbf{v} \right\} d\omega = 0 \tag{21.10}$$

Since Γ is arbitrary, this implies that

$$\frac{\partial \mu}{\partial t} + g_{\alpha,\alpha} = \frac{1}{c^2} d_\alpha v_\alpha, \tag{21.11}$$

where the summation convention is being applied to the Greek indices.

Since the x_α-component of the linear momentum of the particles within the volume element $d\omega$ will be increased at a rate $d_\alpha d\omega$ by the external forces and $\mathbf{g}^{(\alpha)}$ is the current density vector for the flow of this component of momentum, the equation corresponding to (21.9) expressing the conservation of linear momentum is

$$\frac{d}{dt} \int_\Gamma g_\alpha d\omega = - \int_\Sigma \mathbf{g}^{(\alpha)} \cdot \mathbf{n} \, d\sigma + \int_\Gamma d_\alpha d\omega \tag{21.12}$$

Again, by application of the divergence theorem, we can show that this implies that

$$\frac{\partial g_\alpha}{\partial t} + g_{\alpha\beta,\beta} = d_\alpha \tag{21.13}$$

These equations (21.11), (21.13), of conservation of inertial mass (or energy) and linear momentum can be expressed in four-dimensional form by the introduction of Minkowski coordinates x_i and by the definition of a 4-tensor T_{ij} according to the equations

$$T_{\alpha\beta} = g_{\alpha\beta}, \qquad T_{\alpha 4} = T_{4\alpha} = icg_\alpha, \qquad T_{44} = -c^2 \mu \tag{21.14}$$

It may be verified that, with this notation, the equations reduce to the form

$$T_{ij,j} = D_i \tag{21.15}$$

where the 4-force density D_i is given by equation (21.8). T_{ij} is called the *energy–momentum tensor* for the mass–energy distribution. By assuming T_{ij} behaves like a 4-tensor on transformation between inertial frames, equation (21.15) is guaranteed to be valid in all such frames and the special principle of relativity is satisfied.

In the special case of an incoherent cloud of particles flowing with velocity $\mathbf{v}$, we have $g_\alpha = \mu v_\alpha$ and $g_{\alpha\beta} = g_\alpha v_\beta = \mu v_\alpha v_\beta$. Equations (21.14) now yield

$$T_{ij} = \mu_{00} V_i V_j \tag{21.16}$$

where $\mathbf{V} = (1 - v^2/c^2)^{-1/2} (\mathbf{v}, ic)$ is the 4- velocity of flow and

$$\mu_{00} = (1 - v^2/c^2)\mu \tag{21.17}$$

Since the density of rest mass of the cloud observed from S is $\mu \sqrt{(1 - v^2/c^2)}$, the rest mass of the particles in the volume element $d\omega$ is $\mu \sqrt{(1 - v^2/c^2)} d\omega$. Observed from a frame S_0 in which the particles in this element are momentarily stationary, the volume of the element will be $d\omega_0$; hence, the density of proper mass in S_0 is $\mu \sqrt{(1 - v^2/c^2)} d\omega/d\omega_0 = \mu(1 - v^2/c^2) = \mu_{00}$, having used equation

(21.5). μ_{00} is accordingly referred to as the *proper density of proper mass* of the cloud; it is a 4-invariant and equation (21.16) clearly expresses T_{ij} as a 4-tensor.

In many circumstances, the 4-force density D_i of the external force field can be expressed as the divergence of a second rank tensor, i.e. we can write

$$D_i = -S_{ij,j} \tag{21.18}$$

Equation (21.15) then reduces to

$$(T_{ij} + S_{ij})_{,j} = 0 \tag{21.19}$$

This shows that the external force field can be treated as an additional component of the original mass–energy distribution, contributing its own energy–momentum tensor S_{ij}. The field is then regarded as possessing energy of density $c^2\mu = -S_{44}$, x_α-component of momentum of density $g_\alpha = S_{\alpha4}/ic$, and the momentum flow within the field is described by $g_{\alpha\beta} = S_{\alpha\beta}$. There being no external forces operating on the enlarged system, it is said to be *isolated*, and if T_{ij} is taken to denote the combined energy–momentum tensor, the equations of conservation of energy and momentum for the overall distribution take the form

$$T_{ij,j} = 0 \tag{21.20}$$

i.e. T_{ij} has a vanishing divergence.

22. Energy–momentum tensor for a fluid

In this section we will calculate the equations of motion for an elastic fluid moving under the action of an external force field of density d_α. From these, the energy–momentum tensor for the fluid, including its internal stress field, can be derived.

Let $\tau_{\beta\alpha}$ be the stress tensor, i.e. $(\tau_{1\alpha}, \tau_{2\alpha}, \tau_{3\alpha})$ are the components of the force exerted across unit area, whose normal is parallel to the x_α-axis, by the particles on the side for which x_α takes lesser values upon the particles on the side for which x_α takes greater values. Consider a fluid element in the shape of a small tetrahedron, three of whose faces are normal to the axes and whose fourth face has unit normal n_α (Fig. 4). If $\delta x_1, \delta x_2, \delta x_3$ are the respective lengths of the edges parallel to the axes, the stress force acting on the face parallel to the coordinate plane Ox_2x_3 will have components $\frac{1}{2}(\tau_{11}, \tau_{21}, \tau_{31})\delta x_2\delta x_3$; the forces acting on the two faces parallel to the other two coordinate planes can be calculated similarly. Let $s\delta A$ be the force applied to the sloping face, δA being its area. Then the x_α-component of the equation of motion of the element is

$$\frac{1}{2}\tau_{\alpha1}\delta x_2\delta x_3 + \frac{1}{2}\tau_{\alpha2}\delta x_3\delta x_1 + \frac{1}{2}\tau_{\alpha3}\delta x_1\delta x_2 + s_\alpha\delta A + \frac{1}{6}\delta x_1\delta x_2\delta x_3 d_\alpha = \frac{dp_\alpha}{dt} \tag{22.1}$$

where p_α is the momentum of the element. Since the face of area $\frac{1}{2}\delta x_2\delta x_3$ is the projection of the area δA on to the coordinate plane Ox_2x_3, $n_1\delta A = \frac{1}{2}\delta x_2\delta x_3$;

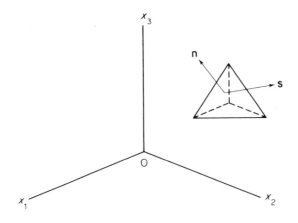

FIG. 4

similarly, $n_2\delta A = \frac{1}{2}\delta x_3 \delta x_1$, $n_3\delta A = \frac{1}{2}\delta x_1 \delta x_2$. Thus,

$$s_\alpha = -(\tau_{\alpha 1} n_1 + \tau_{\alpha 2} n_2 + \tau_{\alpha 3} n_3) + \text{(terms of third order in } \delta x_\alpha)/\delta A. \quad (22.2)$$

In the limit as $\delta x_\alpha \to 0$, this gives

$$s_\alpha = -\tau_{\alpha\beta} n_\beta \quad (22.3)$$

Now consider the motion of a small element of fluid, of any shape, bounded by a surface Σ (Σ moves with the fluid). If $d\sigma$ is an element of Σ (Fig. 5) whose unit normal is n_α, the force exerted on it by the neighbouring fluid is $-\tau_{\alpha\beta} n_\beta d\sigma$ and the resultant stress force on the complete element is therefore

$$-\int_\Sigma \tau_{\alpha\beta} n_\beta d\sigma = -\int_\Gamma \frac{\partial \tau_{\alpha\beta}}{\partial x_\beta} d\omega \quad (22.4)$$

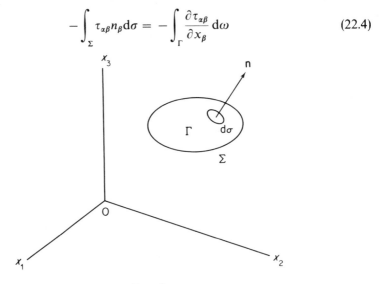

FIG. 5

where Γ is the interior of Σ and we have used the divergence theorem. Thus, the force density for the stress field is $-\tau_{\alpha\beta,\,\beta}$.

Let g_α be the total momentum density for the fluid, including the elastic potential energy generated by the stress field. If $\delta\omega$ now represents the volume of the fluid element inside Σ, the momentum of the element is $g_\alpha\delta\omega$ and the rate of change of momentum is

$$\frac{d}{dt}(g_\alpha\delta\omega) = \frac{dg_\alpha}{dt}\delta\omega + g_\alpha\frac{d}{dt}(\delta\omega) \tag{22.5}$$

where the derivatives are calculated following the fluid motion. Thus,

$$\frac{dg_\alpha}{dt} = \frac{\partial g_\alpha}{\partial t} + v_\beta g_{\alpha,\,\beta} \tag{22.6}$$

where v_α is the velocity of flow. During a short time δt, the surface element $d\sigma$ traces out a volume $\mathbf{v}\cdot\mathbf{n}\,d\sigma\,\delta t$ and the increase in the volume of $\delta\omega$ is accordingly

$$\delta t\int_\Sigma \mathbf{v}\cdot\mathbf{n}\,d\sigma = \delta t\int_\Gamma \operatorname{div}\mathbf{v}\,d\omega = \delta t\,\delta\omega\,v_{\beta,\,\beta} \tag{22.7}$$

Thus,

$$\frac{d}{dt}(\delta\omega) = \delta\omega v_{\beta,\,\beta} \tag{22.8}$$

Equations (22.5), (22.6) now give for the rate of momentum change

$$\left(\frac{\partial g_\alpha}{\partial t} + v_\beta g_{\alpha,\,\beta} + g_\alpha v_{\beta,\,\beta}\right)d\omega = \left(\frac{\partial g_\alpha}{\partial t} + \frac{\partial}{\partial x_\beta}(g_\alpha v_\beta)\right)d\omega \tag{22.9}$$

We can now write down the equation of motion of the element, viz.

$$\left(\frac{\partial g_\alpha}{\partial t} + (g_\alpha v_\beta),_\beta\right)d\omega = (d_\alpha - \tau_{\alpha\beta,\,\beta})d\omega \tag{22.10}$$

or

$$\frac{\partial g_\alpha}{\partial t} + (g_\alpha v_\beta + \tau_{\alpha\beta}),_\beta = d_\alpha \tag{22.11}$$

At this stage it should be noted that we are disregarding any flow of heat which may take place by conduction within the fluid. Such a flow of energy would contribute its own momentum and further terms would need to be included in equation (22.11) to allow for this.

We next calculate the equation of energy for the fluid element. The rate at which the stress force acting upon $d\sigma$ does work is $-\tau_{\alpha\beta}n_\beta v_\alpha d\sigma$ and the total rate of doing work by these forces on the element is therefore

$$-\int_\Sigma v_\alpha\tau_{\alpha\beta}n_\beta d\sigma = -\int_\Gamma \frac{\partial}{\partial x_\beta}(v_\alpha\tau_{\alpha\beta})d\omega \tag{22.12}$$

by the divergence theorem. The rate of doing work by the external forces on the element is $d_\alpha v_\alpha d\omega$. If μ is the density of the total inertial mass of the fluid (including the elastic potential energy and any heat generated by compression), the energy of the element is $c^2 \mu d\omega$. Thus, the equation of energy is

$$\frac{d}{dt}(c^2 \mu d\omega) = \{d_\alpha v_\alpha - (v_\alpha \tau_{\alpha\beta}), {}_\beta\} d\omega \tag{22.13}$$

Using equation (22.8) and

$$\frac{d\mu}{dt} = \frac{\partial \mu}{\partial t} + v_\beta \mu, {}_\beta \tag{22.14}$$

equation (22.13) gives

$$\frac{\partial \mu}{\partial t} + \{\mu v_\beta + \frac{1}{c^2} v_\alpha \tau_{\alpha\beta}\}, {}_\beta = \frac{1}{c^2} d_\alpha v_\alpha \tag{22.15}$$

We now compare the equations (22.11), (22.15) with the equations (21.13), (21.11), for a general mass–energy flow. It is seen that, for the fluid, we must take

$$g_\alpha = \mu v_\alpha + \frac{1}{c^2} v_\beta \tau_{\beta\alpha} \tag{22.16}$$

$$g_{\alpha\beta} = g_\alpha v_\beta + \tau_{\alpha\beta} \tag{22.17}$$

Substituting for g_α from equation (22.16) into (22.17), we get the alternative formula

$$g_{\alpha\beta} = \mu v_\alpha v_\beta + \tau_{\alpha\beta} + \frac{1}{c^2} \tau_{\gamma\alpha} v_\gamma v_\beta \tag{22.18}$$

Equations (21.14) now yield the components of the energy–momentum tensor, viz.

$$\left. \begin{array}{l} T_{\alpha\beta} = \mu v_\alpha v_\beta + \tau_{\alpha\beta} + \dfrac{1}{c^2} \tau_{\gamma\alpha} v_\gamma v_\beta \\[2ex] T_{\alpha 4} = T_{4\alpha} = ic(\mu v_\alpha + \dfrac{1}{c^2} v_\beta \tau_{\beta\alpha}) \\[2ex] T_{44} = -c^2 \mu \end{array} \right\} \tag{22.19}$$

Alternative forms for these components involving the 4-velocity of flow V_i can also be found (see Exercise 68 at the end of this chapter).

Two special cases of these results are of great importance. The first is that of an incoherent cloud for which we have $\tau_{\alpha\beta} = 0$. Then $T_{\alpha\beta} = \mu v_\alpha v_\beta$, $T_{\alpha 4} = ic\mu v_\alpha$, $T_{44} = -c^2 \mu$, and using equation (21.17), we derive equation (21.16) again. It should, however, be noted that equation (21.17) is not valid in the general case of an elastic fluid, since the inertial mass density μ includes a component due to the elastic energy and the value of this component in the rest frame cannot be found by simple multiplication by $(1 - v^2/c^2)$.

The second special case is that of a *perfect fluid*. This is defined to be a fluid in which there are no shearing stresses in a frame S^0 relative to which it is at rest. Thus, in S^0, $\tau^0_{\alpha\beta} = p\delta_{\alpha\beta}$ where p is the *pressure*. Equations (22.19) now give for the components of the energy–momentum tensor in S^0 the values

$$(T^0_{ij}) = \begin{bmatrix} p & 0 & 0 & 0 \\ 0 & p & 0 & 0 \\ 0 & 0 & p & 0 \\ 0 & 0 & 0 & -c^2\mu_{00} \end{bmatrix} \tag{22.20}$$

If V_i is the 4-velocity of flow of the fluid, it is now easy to verify that the tensor equation

$$T_{ij} = (\mu_{00} + p/c^2)V_iV_j + p\delta_{ij} \tag{22.21}$$

is valid in the frame S^0 (μ_{00}, p being invariants) and, hence, is valid in all frames. Thus, if μ is the density of inertial mass in a frame S,

$$\mu = -T_{44}/c^2 = \frac{\mu_{00} + p/c^2}{1 - v^2/c^2} - \frac{p}{c^2} \tag{22.22}$$

or

$$\mu + p/c^2 = (\mu_{00} + p/c^2)/(1 - v^2/c^2) \tag{22.23}$$

It now follows that

$$\begin{aligned} g_\alpha = T_{\alpha4}/ic &= (\mu_{00} + p/c^2)\frac{v_\alpha}{1 - v^2/c^2} \\ &= (\mu + p/c^2)v_\alpha \end{aligned} \tag{22.24}$$

Comparing this last equation with equation (22.16), we deduce that $pv_\alpha = v_\beta\tau_{\beta\alpha}$ identically, which implies that

$$\tau_{\beta\alpha} = p\delta_{\beta\alpha} \tag{22.25}$$

i.e. there is no shearing stress in any frame and the pressure is the same in all frames.

23. Angular momentum

A particle having momentum $p_\alpha = mv_\alpha$ at a point x_α relative to a frame S is defined to have *angular momentum*

$$h_{\alpha\beta} = (x_\alpha - a_\alpha)p_\beta - (x_\beta - a_\beta)p_\alpha \tag{23.1}$$

about the fixed point a_α. Clearly, this defines the angular momentum as an anti-symmetric 3-tensor. In elementary mechanics, a more usual definition is by the vector product $(\mathbf{x} - \mathbf{a}) \times \mathbf{p}$, i.e. the pseudovector $e_{\alpha\beta\gamma}(x_\beta - a_\beta)p_\gamma$; however, the components of this pseudovector are found to be (h_{23}, h_{31}, h_{12}) and these are three of the non-vanishing components of $h_{\alpha\beta}$(the other three are $h_{32} = -h_{23}$, etc.), so that the definitions are essentially equivalent.

If f_α is the 3-force acting on the particle, then

$$\frac{dh_{\alpha\beta}}{dt} = \frac{dx_\alpha}{dt}p_\beta - \frac{dx_\beta}{dt}p_\alpha + (x_\alpha - a_\alpha)\frac{dp_\beta}{dt} - (x_\beta - a_\beta)\frac{dp_\alpha}{dt},$$

$$= (x_\alpha - a_\alpha)f_\beta - (x_\beta - a_\beta)f_\alpha \tag{23.2}$$

since $p_\alpha = mv_\alpha = m\dot{x}_\alpha$ and $\dot{p}_\alpha = f_\alpha$. The right-hand member of the last equation is another anti-symmetric 3-tensor called the *moment* of the force f_α about the point a_α. Denoting this moment by $m_{\alpha\beta}$, we have derived the equation of angular momentum, viz.

$$\frac{dh_{\alpha\beta}}{dt} = m_{\alpha\beta} \tag{23.3}$$

If $m_{\alpha\beta} = 0$, then $h_{\alpha\beta}$ is constant and the angular momentum is conserved.

In the case of the continuous distribution of mass–energy considered in section 21, the momentum of an element of volume $d\omega$ is $g_\alpha d\omega$ and the angular momentum of the whole system about a_α is defined by the equation

$$h_{\alpha\beta} = \int_\Gamma \{(x_\alpha - a_\alpha)g_\beta - (x_\beta - a_\beta)g_\alpha\}d\omega$$

$$= \frac{1}{ic}\int_\Gamma \{(x_\alpha - a_\alpha)T_{\beta4} - (x_\beta - a_\beta)T_{\alpha4}\}d\omega \tag{23.4}$$

where Γ is the region occupied by the system and T_{ij} is the energy–momentum tensor. If the system is imagined to be situated in otherwise empty space, so that there is no container exerting forces on its bounding surface, the region Γ can be extended to include the rest of space. In these circumstances, differentiating equation (23.4) with respect to t, we find

$$\frac{dh_{\alpha\beta}}{dt} = \int_\Gamma \{(x_\alpha - a_\alpha)T_{\beta4,4} - (x_\beta - a_\beta)T_{\beta4,4}\}d\omega$$

$$= \int_\Gamma \{(x_\alpha - a_\alpha)(D_\beta - T_{\beta\gamma,\gamma}) - (x_\beta - a_\beta)(D_\alpha - T_{\alpha\gamma,\gamma})\}d\omega,$$

$$= \int_\Gamma \{(x_\alpha - a_\alpha)d_\beta - (x_\beta - a_\beta)d_\alpha\}d\omega$$

$$- \int_\Gamma \{[(x_\alpha - a_\alpha)T_{\beta\gamma}]_{,\gamma} - [(x_\beta - a_\beta)T_{\alpha\gamma}]_{,\gamma} - x_{\alpha,\gamma}T_{\beta\gamma} + x_{\beta,\gamma}T_{\alpha\gamma}\}d\omega \tag{23.5}$$

having used equations (21.8) and (21.15). Applying the divergence theorem to the first two terms involving the energy–momentum tensor, since T_{ij} vanishes at a great distance from the distribution, these terms make zero contribution. Thus, the equation reduces to the form

$$\frac{dh_{\alpha\beta}}{dt} = \int_\Gamma \{(x_\alpha - a_\alpha)d_\beta - (x_\beta - a_\beta)d_\alpha\}d\omega + \int_\Gamma (T_{\beta\alpha} - T_{\alpha\beta})d\omega \qquad (23.6)$$

(Note that $x_{\alpha,\gamma} = \delta_{\alpha\gamma}$.) The first integral in the right-hand member of equation (23.6) is the moment $m_{\alpha\beta}$ of the external field forces about a_α and it follows that the equation of angular momentum (23.3) is valid for the distribution if $T_{\alpha\beta}$ is symmetric. Since $T_{\alpha4} = T_{4\alpha}$, this condition is equivalent to the requirement that the energy–momentum tensor should be symmetric.

The assumption that T_{ij} is symmetric, and hence that the angular momentum of a continuous distribution not acted upon by external forces is conserved, is always made and is, indeed, necessary for the development of the general theory of relativity (see section 47). Reference to equations (22.19) indicates that this assumption implies that the stress tensor $\tau_{\alpha\beta}$ for an elastic fluid cannot be symmetric as was always assumed in the classical theory; it will, however, be very nearly symmetric in the case when its components and the flow velocity are sufficiently small.

Exercises 3

1. Obtain the transformation equations for v by differentiating the Lorentz transformation.

2. Obtain the transformation equations for the acceleration a by differentiating the transformation equations for v and express them in the form

$$\bar{a}_x = \frac{(1 - u^2/c^2)^{3/2}}{(1 - v_x u/c^2)^3} a_x$$

$$\bar{a}_y = \frac{1 - u^2/c^2}{(1 - v_x u/c^2)^2} \left(a_y + \frac{v_y u/c^2}{1 - v_x u/c^2} a_x \right)$$

$$\bar{a}_z = \frac{1 - u^2/c^2}{(1 - v_x u/c^2)^2} \left(a_z + \frac{v_z u/c^2}{1 - v_x u/c^2} a_x \right)$$

Deduce that a point which has uniform acceleration in one inertial frame has not, in general, uniform acceleration in another.

3. A nucleus is moving along a straight line when it emits an electron. As seen from the nucleus, the electron's velocity is $6c/7$ making an angle of $60°$ with its direction of motion. A stationary observer measures the angle between the lines of motion of nucleus and electron to be $30°$. Calculate the speed of the nucleus. (Ans. $3c/5$.)

4. A nucleus is moving with velocity $3c/5$ when it emits a β-particle with velocity $3c/4$ relative to itself in a direction perpendicular to its line of motion. Calculate the velocity and direction of motion of the β-particle as seen by a stationary observer. If the β-particle is emitted with velocity $3c/4$ in such a

direction that the stationary observer sees its line of motion to be perpendicular to that of the nucleus, calculate the direction of emission as seen from that nucleus and the velocity of the β-particle as seen by the stationary observer. (Ans. $3c/5$ at $45°$ to line of motion; $\pi - \alpha$ to line of motion where $\cos \alpha = 4/5$; $9c/16$.)

5. Show that the 4-velocity $\mathbf{V}$ is of constant magnitude ic.

6. A beam of light is being propagated in the xy-plane of S at an angle α to the x-axis. Relative to $\bar{S}$ it is observed to make an angle $\bar{\alpha}$ with $\overline{Ox}$. Prove the aberration of light formula, viz.

$$\cot \bar{\alpha} = \frac{\cot \alpha - (u/c)\operatorname{cosec} \alpha}{(1 - u^2/c^2)^{1/2}}$$

Deduce that, if $u \ll c$, then

$$\Delta\alpha = \bar{\alpha} - \alpha = \frac{u}{c} \sin \alpha$$

approximately.

7. A particle of rest mass m_0 is moving under the action of a force $\mathbf{f}$ with velocity $\mathbf{v}$. Show that

$$\mathbf{f} = \frac{m_0}{(1 - v^2/c^2)^{1/2}} \frac{d\mathbf{v}}{dt} + \frac{m_0 v\dot{v}/c^2}{(1 - v^2/c^2)^{3/2}} \mathbf{v}$$

Hence, if the acceleration $d\mathbf{v}/dt$ is parallel to $\mathbf{v}$, show that

$$\mathbf{f} = \frac{m_0}{(1 - v^2/c^2)^{3/2}} \frac{d\mathbf{v}}{dt},$$

and if the acceleration is perpendicular to $\mathbf{v}$, then

$$\mathbf{f} = \frac{m_0}{(1 - v^2/c^2)^{1/2}} \frac{d\mathbf{v}}{dt}$$

8. Two particles are moving along the x-axis of a frame S with velocities v_1, v_2. Calculate the velocity u with which a parallel frame $\bar{S}$ must move parallel to the x-axis of S, if the particles have equal and opposite velocities relative to S. Show that the magnitude of these velocities is

$$\frac{c^2 - v_1 v_2 - (c^2 - v_1^2)^{1/2}(c^2 - v_2^2)^{1/2}}{v_1 - v_2}$$

assuming $v_1 > v_2 > 0$.

9. A bullet of length d is moving with velocity v. The line of sight from a camera makes an angle α with the bullet's velocity. Behind the bullet and parallel to its axis is a stationary measuring scale. If the camera takes a photograph of the bullet against the background provided by the scale, show that the bullet's length as it appears on the scale is

$$\frac{(1 - v^2/c^2)^{1/2}d}{1 + v \cos \alpha/c}.$$

10. A cart rolls on a table with velocity kc. A smaller cart rolls on the first in the same direction with velocity kc relative to the first cart. A third cart rolls on the second with relative velocity kc, and so on up to n carts. If cv_r is the velocity of the rth cart relative to the table, prove that

$$v_{r+1} = \frac{v_r + k}{1 + kv_r}$$

Deduce that

$$v_n = \frac{(1+k)^n - (1-k)^n}{(1+k)^n + (1-k)^n}$$

What is the limit of v_n as $n \to \infty$? (Ans. c.)

11. A nucleus disintegrates into two parts, A and B which move with equal and opposite velocities of magnitude V. A then ejects an electron whose velocity observed from A has magnitude V and direction perpendicular to the direction of A's motion. Show that, as observed from B, the electron's velocity makes an angle α with the direction of A's motion, where $\tan \alpha = \frac{1}{2}(1 - V^2/c^2)$ and calculate the magnitude of the velocity of the electron relative to B. (Ans. $V\{4 + (1 - V^2/c^2)^2\}^{1/2}/(1 - V^2/c^2)$.)

12. A rocket moves along the x-axis in S, commencing its motion with velocity v_0 and ending it with velocity v_1. If w is the jet velocity as measured by the crew (assumed constant), show that the mass ratio of the manoeuvre (i.e. initial mass/final mass) as measured by the crew is

$$\left[\frac{(c + v_1)(c - v_0)}{(c - v_1)(c + v_0)} \right]^{c/2w}$$

What does this reduce to as $c \to \infty$? Deduce that, if the rocket starts from rest in S and its jet is a stream of photons, the mass ratio to velocity v is

$$\sqrt{\left(\frac{c+v}{c-v} \right)}$$

Show that, with a mass ratio of 6, the rocket can attain 35/37 of the velocity of light in S.

13. $S, \bar{S}, \bar{\bar{S}}$ are inertial frames with their axes parallel. $\bar{\bar{S}}$ has a velocity u relative to $\bar{S}$ and $\bar{S}$ has a velocity v relative to S, both velocities being parallel to the x-axes. If transformation from $\bar{\bar{S}}$ to $\bar{S}$ involves a rotation through an angle α of the axes in space-time and transformation from S to $\bar{S}$ a rotation β, a transformation from S to $\bar{\bar{S}}$ involves a rotation γ where $\gamma = \alpha + \beta$. Deduce from this equation the relativistic law for the composition of velocities, viz.

$$w = \frac{u + v}{1 + uv/c^2}$$

14. A force **f** acts upon a particle of mass m whose velocity is **v**. Show that

$$\mathbf{f} = m\frac{d\mathbf{v}}{dt} + \frac{\mathbf{f}\cdot\mathbf{v}}{c^2}\mathbf{v}$$

15. An electrified particle having charge e and rest mass m_0 moves in a uniform electric field of intensity E parallel to the x-axis. If it is initially at rest at the origin, show that it moves along the x-axis so that at time t

$$x = \frac{c^2}{k}\left\{\sqrt{\left(1 + \frac{k^2}{c^2}t^2\right)} - 1\right\}$$

where $k = eE/m_0$. Show that this motion approaches that predicted by classical mechanics as $c \to \infty$. (It may be assumed that the force acting upon the particle is eE in the direction of the field at all times.)

16. A tachyon transmitter always emits a tachyon at a speed $v > c$ relative to itself. Observers A, B are equipped with such transmitters and B is moving away from A with velocity $u < c$. A transmits a tachyon towards B, who is at a distance d as measured from A when he receives it. B immediately transmits a tachyon back towards A. Show that A receives this tachyon a time

$$\frac{d}{v(v-u)}(2u - v - u^2v/c^2)$$

after transmitting his own. Deduce that A receives the reply before (!) his act of transmission if

$$\frac{v}{c} > \frac{c + \sqrt{(c^2 - u^2)}}{u}$$

17. **v**, **V** are the 3- and 4-velocities of a point. If $\mathbf{a} = d\mathbf{v}/dt$ is the 3-acceleration and $\mathbf{A} = d\mathbf{V}/d\tau$ is the 4-acceleration, prove that

$$A^2 = (1 - v^2/c^2)^{-3}\{(c^2 - v^2)a^2 + 2v\dot{v}\mathbf{a}\cdot\mathbf{v} - v^2\dot{v}^2\}/c^2$$

18. A mirror moves perpendicular to its plane with velocity v and away from a source of light. A ray from the source is reflected by the mirror. If θ is the ray's angle of incidence, show that the angle of reflection is ϕ, where

$$\cos\phi = \frac{(v^2 + c^2)\cos\theta - 2cv}{v^2 + c^2 - 2vc\cos\theta}$$

19. Two trains, each having the same rest length L, are moving in opposite directions with equal speeds U on parallel tracks. State the time T they take to pass one another according to a classical, non-relativistic, calculation. Show that the time taken, as measured by a driver of one of the trains and using a relativistic calculation, is also T.

20. **v**, $\bar{\mathbf{v}}$ are the velocities of a point relative to the inertial frames $S, \bar{S}$ respectively. Representing these vectors as position vectors in an independent $\mathscr{E}_3$,

show that

$$\beta \bar{\mathbf{v}} = Q\left[\mathbf{v} + \mathbf{u}\left\{\frac{\mathbf{u} \cdot \mathbf{v}}{u^2}(\beta - 1) + \beta\right\}\right]$$

where $\beta = (1 - u^2/c^2)^{-1/2}$ and $Q = 1/(1 + \mathbf{u} \cdot \mathbf{v}/c^2)$.
Show further that

$$u^2 \beta \bar{\mathbf{v}} = Q[(1 - \beta)\mathbf{u} \times (\mathbf{v} \times \mathbf{u}) + \beta u^2(\mathbf{u} + \mathbf{v})]$$

and hence verify that

$$\bar{v}^2 = Q^2[(\mathbf{u} + \mathbf{v})^2 - (\mathbf{v} \times \mathbf{u})^2/c^2]$$

21. A luminous disc of radius a has its centre fixed at the point $(\bar{x}, 0, 0)$ of the $\bar{S}$-frame and its plane is perpendicular to the $\bar{x}$-axis. It is observed from the origin in the S-frame at the instant the origins of the two frames coincide and is measured to subtend an angle 2α. Prove that, if $a \ll \bar{x}$, then

$$\tan\alpha = \frac{a}{\bar{x}}\bigg/\sqrt{\left(\frac{c+u}{c-u}\right)}$$

(*Hint*: employ the aberration of light formula, exercise 6 above.)

22. A particle moves along the x-axis of the frame S with velocity v and acceleration a. Show that the particle's acceleration in $\bar{S}$ is

$$\bar{a} = \frac{(1 - u^2/c^2)^{3/2}}{(1 - uv/c^2)^3} a$$

If the particle always has constant acceleration α relative to an inertial frame in which it is instantaneously at rest, prove that

$$\frac{d}{dt}(\beta v) = \alpha$$

where $\beta = (1 - v^2/c^2)^{-1/2}$ and t is time in S.

Assuming that the particle is at rest at the origin of S at $t = 0$, show that its x-coordinate at time t is given by

$$\alpha x = c^2[(1 + \alpha^2 t^2/c^2)^{1/2} - 1]$$

23. Three rectangular Cartesian inertial frames S, $\bar{S}$, $\bar{\bar{S}}$ are initially coincident. As seen from S, $\bar{S}$ moves with velocity u parallel to Ox and, as seen from $\bar{S}$, $\bar{\bar{S}}$ moves with velocity v parallel to $\bar{O}\bar{y}$. If the direction of S's motion as seen from S makes an angle θ with Ox and the direction of S's motion as seen from $\bar{\bar{S}}$ makes an angle ϕ with $\bar{\bar{O}}\bar{\bar{x}}$, prove that

$$\tan\theta = \frac{v}{u}\left(1 - \frac{u^2}{c^2}\right)^{1/2}, \quad \tan\phi = \frac{v}{u}\left(1 - \frac{v^2}{c^2}\right)^{-1/2}$$

Deduce that, if $u, v \ll c$, then

$$\phi - \theta = uv/2c^2$$

approximately.

24. The inertial frames S, $\overline{S}$ have their axes parallel and the origin of $\overline{S}$ moves along the x-axis of S with velocity u. A rigid rod lies along the $\overline{x}$-axis of $\overline{S}$ and is attached to it. If $\overline{l}$ is the rod's length as measured in $\overline{S}$ and l is its length measured in S, show that $l = \overline{l}(1 - u^2/c^2)^{1/2}$. S' is a third parallel inertial frame whose origin also moves along the x-axis of S. Observed from S', the origins of S and $\overline{S}$ have equal and opposite velocities. Show that the velocity of S' observed from S is

$$\frac{c^2}{u}\left[1 - (1 - u^2/c^2)^{1/2}\right]$$

A rod, identical to the one already referred to, lies along the x'-axis of S' and moves with this frame. Its length observed from S is L. Show that

$$L = \left(\frac{2l}{l+\overline{l}}\right)^{1/2}\overline{l}$$

25. Two particles, each having rest mass m_0, are moving in perpendicular directions with the same speed $\frac{1}{2}c$. They collide and cohere to form a single particle. Show that its rest mass is $\sqrt{(14/3)}m_0$. (Assume there is no radiation of energy.)

26. A particle of rest mass m_1 and speed v collides with a particle of rest mass m_2 which is stationary. After collision the two particles coalesce. Assuming that there is no radiation of energy, show that the rest mass of the combined particle is M, where

$$M^2 = m_1^2 + m_2^2 + \frac{2m_1 m_2}{(1 - v^2/c^2)^{1/2}}$$

and find its speed.

27. A particle is moving with velocity u when it collides with a stationary particle having the same rest mass. After the collision the particles are moving at angles θ, ϕ with the direction of motion of the first particle before collision. Show that

$$\tan\theta\tan\phi = \frac{2}{\gamma + 1}$$

where $\gamma = (1 - u^2/c^2)^{-1/2}$. (If $c \to \infty$, $\gamma \to 1$ and $\theta + \phi = \frac{1}{2}\pi$. This is the prediction of classical mechanics. However, if the particles are electrons and u is near to c in value, $\theta + \phi < \frac{1}{2}\pi$. This effect has been observed in a Wilson cloud chamber.) (*Hint:* Refer the collision to an inertial frame in which both particles have equal and opposite velocities prior to collision.)

28. A body of mass M disintegrates while at rest into two parts of rest masses M_1 and M_2. Show that the energies E_1, E_2 of the parts are given by

$$E_1 = c^2 \frac{M^2 + M_1^2 - M_2^2}{2M}, \quad E_2 = c^2 \frac{M^2 - M_1^2 + M_2^2}{2M}$$

29. Two particles having rest masses m_1, m_2 are moving with velocities u_1, u_2 respectively, when they collide and cohere. If α is the angle between their lines of motion before collision, show that the rest mass of the combined particle is m, where

$$m^2 = m_1^2 + m_2^2 + \frac{2m_1 m_2 (c^2 - u_1 u_2 \cos\alpha)}{\sqrt{\{(c^2 - u_1^2)(c^2 - u_2^2)\}}}.$$

Show that, for all values of α, $m \geqslant m_1 + m_2$ and explain the increase in rest mass.

30. A photon having energy E collides with a stationary electron whose rest mass is m_0. As a result of the collision the direction of the photon's motion is deflected through an angle θ and its energy is reduced to E'. Prove that

$$m_0 c^2 \left(\frac{1}{E'} - \frac{1}{E} \right) = 1 - \cos\theta$$

Deduce that the wavelength λ of the photon is increased by

$$\Delta\lambda = \frac{2h}{m_0 c} \sin^2 \tfrac{1}{2}\theta,$$

where h is Planck's constant. (This is the *Compton effect*. For a photon, take $\lambda = hc/E$.)

31. A particle P having rest mass $2m_0$ collides with a stationary particle Q having rest mass m_0. After the collision, the rest mass of Q is unchanged, but the rest mass of P has been reduced to m_0. If the lines of motion of the two particles after the collision both make an angle of $30°$ with the original line of motion P, calculate the original velocity of P and the momentum acquired by Q. (Ans. $3\sqrt{5}c/7$; $\sqrt{15}m_0 c$.)

32. A particle is moving with velocity v when it disintegrates into two photons having energies E_1, E_2, moving in directions making angles α, β with the original direction of motion and on opposite sides of this direction. Show that

$$\tan\tfrac{1}{2}\alpha \tan\tfrac{1}{2}\beta = \frac{c - v}{c + v}$$

Deduce that, if a photon disintegrates into two photons, they must both move in the same direction as the original photon.

33. A stationary particle having rest mass $3m_0$ disintegrates into a pair of particles, each of rest mass m_0, and a neutrino. The directions of motion of the particle pair are at an angle 2θ, where $\cos\theta = 1/3$. Calculate the energy of the neutrino and show that the speed of each of the other two particles is $3c/5$. (Ans. $\tfrac{1}{2}m_0 c^2$.)

34. A cosmic ray particle has rest mass m_0 and is moving with velocity $3c/5$ relative to a stationary observer. It is seen by this observer to emit a gamma ray photon with energy $m_0 c^2/4$ in a direction making an angle of $60°$ with its original line of motion. Show that the rest mass of the particle is reduced by a quarter and calculate the angle through which its velocity is deflected and its new speed. (Ans. $\tan^{-1} (\sqrt{3}/5)$; $\sqrt{7}c/4$.)

35. A neutron having rest mass m_N is stationary when it disintegrates into a proton (rest mass m_P), an electron (rest mass m_E) and a neutrino. The proton moves in the opposite direction to the other two particles, which move along the same straight line. If T is the kinetic energy of the proton, prove that the kinetic energy of the electron is $c(m_E c - k)^2/2k$, where

$$k = (m_N - m_P)c - \frac{T}{c} - \sqrt{(2m_P T + T^2/c^2)}$$

36. A particle having rest mass m_0 is at rest when it emits two photons, each of energy $\frac{1}{4}m_0 c^2$. The particle recoils with rest mass $\frac{1}{4}m_0$ along a line bisecting the angle between the tracks of the photons. Calculate the angle between these tracks and the particle's velocity of recoil. If the photons are observed by an observer moving with the particle, show that the angle between their tracks is seen to be 2α, where $\sin\alpha = 1/7$. (Ans. $30°$; $\sqrt{3}c/2$.)

37. A nucleus has rest mass M. Whilst at rest, it emits a photon. If the internal energy of the nucleus is reduced by E_0 in the process, show that the energy of the photon is E, where $E = E_0(1 - E_0/2Mc^2)$.

38. The lines of motion of a particle having rest mass m_0 and a photon are perpendicular to one another. The total energies of the particle and photon are E, $\overline{E}$ respectively. If the particle absorbs the photon, show that its rest mass is increased to M_0, where $M_0^2 = m_0^2 + 2E\overline{E}/c^4$.

39. A photon having energy E is moving along the x-axis when it encounters a stationary particle having rest mass m_0. The particle absorbs the photon and then emits another photon having the same energy in a direction parallel to the y-axis. Calculate the direction and magnitude of the final momentum of the particle and show that its rest mass is reduced to the value $\sqrt{(m_0^2 - 2E^2/c^4)}$.

40. A mass $3\lambda m(\lambda > 1)$ at rest disintegrates into three fragments (each of rest mass m) which move apart in directions making equal angles with each other. Show that, in a frame of reference in which one of the fragments is at rest, the angle between the directions of motion of the other two fragments is $2\cot^{-1} (\sqrt{3}\lambda)$.

41. A positron travelling with velocity $3c/5$ is annihilated in a collision with a stationary electron, yielding two photons which emerge in opposite directions along the track of the incoming particle. If m is the rest mass of the electron and positron, show that the photons have energies $3mc^2/4$ and $3mc^2/2$.

42. A positron having momentum p collides with a stationary electron. Both particles are annihilated and two photons are generated whose lines of motion make equal angles α on opposite sides of the line of motion of the positron. Prove

that $p \sin \alpha \tan \alpha = 2mc$ where m is the rest mass of both the positron and electron. If $\alpha = 60°$, calculate the velocity of the positron. (Ans. $4c/5$.)

43. A particle of rest mass m_0 collides elastically with an identical stationary particle and, as a result, its motion is deflected through an angle θ. If T is its KE before the collision and T' is its KE afterwards, show that

$$T' = \frac{T \cos^2 \theta}{1 + T \sin^2 \theta / 2m_0 c^2}$$

44. A particle of rest mass m_1 collides elastically with a stationary particle of rest mass m_2 ($< m_1$) and, as a result, is deflected through an angle θ. If E, E' are the total energies of the particle m_1 before and after collision respectively, prove that

$$\cos \theta = \frac{(E + m_2 c^2)E' - m_2 c^2 E - m_1^2 c^4}{[(E^2 - m_1^2 c^4)(E'^2 - m_1^2 c^4)]^{1/2}}$$

45. A pion having rest mass m_0 is moving along the x-axis of an inertial frame $Oxyz$ with speed $4c/5$, when it disintegrates into a muon having rest mass $2m_0/3$ and a neutrino. The neutrino moves parallel to the y-axis. Prove that the angle made by the muon's velocity with the x-axis is $\tan^{-1}(1/8)$ and calculate the energy of the neutrino. (Ans. $m_0 c^2/6$.)

46. A nucleus having rest mass m_0 disintegrates when at rest into a pair of identical fragments of rest mass $\frac{1}{2}\lambda m_0$ ($\lambda < 1$). Show that the speed of one particle relative to the other is $2\sqrt{(1 - \lambda^2)}c/(2 - \lambda^2)$. If λ is small, show that this speed is less than c by a fraction $\lambda^4/8$.

47. A positron collides with a stationary electron and the two particles are annihilated. Two photons are generated, the lines of motion of which make angles of $30°$ and $90°$ with the original line of motion of the positron. Calculate the original velocity of the positron and show that the energy of one of the photons is equal to the internal energy of the electron. (The rest masses of an electron and a positron are equal.) (Ans. $\sqrt{3}c/2$.)

48. A moving positron collides with a stationary electron. Both particles are annihilated and two photons are generated, the lines of motion of which both make angles of $60°$ with the original line of motion of the positron. Prove that the total energy of each photon is $4m_0 c^2/3$, where m_0 is the rest mass of the positron and of the electron.

49. A nucleus having rest mass m_0 is moving with velocity $4c/5$ when it emits a photon having energy $m_0 c^2/3$ in a direction making an angle of $60°$ with the line of motion of the nucleus. Show that the subsequent direction of motion of the nucleus makes an angle $\tan^{-1}(\sqrt{3}/7)$ with its initial direction of motion and calculate the new rest mass of the nucleus. Show that, relative to an inertial frame in which the nucleus was initially at rest, the line of motion of the photon makes an angle of $120°$ with the original direction of motion of the nucleus. (Ans. $\sqrt{13}m_0/6$.)

50. A pion has rest mass m_0 and momentum p when it disintegrates into a pair of photons having energies E and E'. The directions of motion of the photons are perpendicular and that of the photon having energy E makes an angle α with the original direction of motion of the pion. Prove that

$$p = m_0 c (\sin 2\alpha)^{-1/2}, \quad E = m_0 c^2 \sqrt{(\tfrac{1}{2}\cot \alpha)},$$
$$E' = m_0 c^2 \sqrt{(\tfrac{1}{2}\tan \alpha)}.$$

51. A particle having rest mass m_0 is moving with an unknown velocity when it absorbs a neutrino whose energy is $3m_0 c^2/2$. The angle between the paths of the particle and neutrino is α, where $\cos \alpha = 1/3$. After absorption, the rest mass of the particle is $2m_0$. Calculate the original velocity of the particle and show that its path is deflected through an angle β, where $\tan \beta = 4\sqrt{2/5}$, as a result of the encounter. (Ans. $3c/5$.)

52. A particle whose rest mass is m_0 moves along the x-axis of an inertial frame under the action of a force

$$f = \frac{2m_0 c^2 a}{(a-x)^2}$$

At time $t = 0$, the particle is at rest at the origin O. Show that the time taken for the particle to move from O to a point x ($< a$) is given by

$$t = \frac{1}{3c}\left[\frac{x}{a}\right]^{1/2} (x + 3a)$$

53. Oxy are rectangular axes of an inertial frame. A particle having rest mass m_0 is projected from the origin with momentum p_0 along Ox. It is acted upon by a constant force f parallel to Oy. Show that its path is the catenary

$$y = \frac{w_0}{f}\cosh\left(\frac{fx}{cp_0} - 1\right)$$

where $w_0^2 = m_0^2 c^4 + p_0^2 c^2$.

54. A particle having rest mass m_0 moves along a straight line under the action of a frictional force of magnitude $m_0 v/k$ opposing its motion; v is the speed of the particle and k is a constant. Show that the time which elapses whilst the particle's velocity is reduced from $4c/5$ to $3c/5$ is $[\log (3/2) + 5/12]k$.

55. A particle having rest mass m_0 moves on the x-axis under an attractive force to the origin of magnitude $2m_0 c^2/x^2$. Initially it is at rest at $x = 2$. Show that its motion is simple harmonic with period $4\pi/c$.

56. A space ship, with its motors closed down, is moving at high velocity v through stationary interstellar gas which causes a retardation as measured by the crew of magnitude αv^2. Show that the distance it moves through the gas whilst its velocity is reduced from V to U is

$$\frac{1}{\alpha}\left|\frac{1}{x} - \tfrac{1}{2}\log\frac{1+x}{1-x}\right|_U^V$$

where $x = (1 - v^2/c^2)^{1/2}$.

57. A particle has rest mass m_0 and 4-momentum $\mathbf{P}$. An observer has 4-velocity $\mathbf{V}$ in the same frame. Show that, for this observer, the particle's:
(i) energy is $-\mathbf{P} \cdot \mathbf{V}$;
(ii) momentum is of magnitude $\sqrt{[\mathbf{P}^2 + (\mathbf{P} \cdot \mathbf{V})^2/c^2]}$;
(iii) velocity is of magnitude $\sqrt{[1 + c^2\mathbf{P}^2/(\mathbf{P} \cdot \mathbf{V})^2]}$.
(*Hint*: All these expressions are invariant.)

58. A particle has rest mass m_0 and moves along the x-axis under the action of a force given at any point having coordinate x by

$$f = -\frac{m_0 c^3 \omega^2 x}{(c^2 - \omega^2 a^2 + \omega^2 x^2)^{3/2}}$$

ω and a being constants. It is projected from the origin with velocity ωa. Show that its velocity at any later time is given by $v^2 = \omega^2(a^2 - x^2)$. What does this imply for the particle's motion?

59. A particle of rest mass m_0 moves along the x-axis of an inertial frame under the action of a force

$$f = \frac{m_0 c^2}{2(1 + 2x^{1/2})^{3/2}}$$

At time $t = 0$, the particle is at rest at the origin. Show that, at any later time t, its coordinate is given by

$$x = 2 + ct - 2\sqrt{(1 + ct)}$$

60. A particle of rest mass m_0 moves under the action of a central force. (r, θ) are its polar coordinates in its plane of motion relative to the force centre as pole. $V(r)$ is its potential energy when at a distance r from the centre. Obtain Lagrange's equations for the motion in the form

$$\frac{d}{dt}(\gamma \dot{r}) - \gamma r \dot{\theta}^2 + \frac{1}{m_0} V' = 0, \quad \frac{d}{dt}(\gamma r^2 \dot{\theta}) = 0$$

where $\gamma = [1 - (\dot{r}^2 + r^2 \dot{\theta}^2)/c^2]^{-1/2}$. Write down the energy equation for the motion and obtain the differential equation for the orbit in the form

$$h^2 u^2 \left(\frac{d^2 u}{d\theta^2} + u\right) = \frac{C - V}{m_0^2 c^2} V'$$

where $u = 1/r$ and h, C are constants. In the inverse square law case when $V = -\mu/r$, deduce that the polar equation of the orbit can be written

$$lu = 1 + e \cos \eta \theta$$

where $\eta^2 = 1 - \mu^2/m_0^2 h^2 c^2$. If $\mu/m_0 hc$ is small, show that the orbit is approximately an ellipse whose major axis rotates through an angle $\pi \mu^2 / m_0^2 h^2 c^2$ per revolution.

61. A particle, having rest mass m_0, is at rest at the origin of the x-axis at time $t = 0$. It is acted upon by a force f, directed along the positive x-axis, whose

magnitude when the particle's velocity is v is given by $f = m_0 kc^2/v$. Show that at time $t(\, > 0)$, $v = c\sin\theta$, where θ is positive acute and $\sec\theta = 1 + kt$. Deduce that, at the same time, the coordinate of the particle is given by $x = c(\tan\theta - \theta)/k$.

62. If $\mathbf{v}$ is the 3-velocity of a particle and $\beta = (1 - v^2/c^2)^{-1/2}$, prove that $\mathbf{v}\cdot\dot{\mathbf{v}} = v\dot{v}$ and

$$\mathbf{v}\cdot\frac{\mathrm{d}}{\mathrm{d}t}(\beta\mathbf{v}) = \beta^3 v\dot{v}$$

If m_0 is the particle's rest mass, define the 3-force $\mathbf{f}$ acting upon it and deduce from the above result that $\mathbf{v}\cdot\mathbf{f} = \dot{m}c^2$, where m is the inertial mass.

63. A particle having rest mass m_0 moves along the x-axis under a force of attraction towards the origin $-m_0\omega^2 x$. It is initially at rest at the point $x = a$. Show that the velocity with which it passes through the origin is

$$\frac{\omega ac\sqrt{(4c^2 + \omega^2 a^2)}}{2c^2 + \omega^2 a^2}$$

64. If the force $\mathbf{f}$ always acts along a normal to a particle's path, show that the speed v of the particle is constant. Write down the equation of motion of the particle and deduce that the curvature of the path is given by $\kappa = f/mv^2$. If the particle moves in a circle of radius a under a constant radial force f, show that its speed v is given by

$$v^2 = 2c^2\lambda[\,\sqrt{(\lambda^2 + 1)} - \lambda]$$

where $\lambda = fa/2m_0c^2$ and m_0 is the particle's rest mass.

65. A nucleus is moving along a straight line when it ejects an electron. As measured by a stationary observer, the speed of the electron is $\tfrac{1}{2}c$ and the angle between the lines of motion of the nucleus and electron is $60°$. If the speed of the electron relative to the nucleus is also $\tfrac{1}{2}c$, calculate the speed of the nucleus. (Ans. $8c/17$.)

66. A particle having rest mass m_0, initially at rest at the origin of an inertial frame, moves along its x-axis under the action of a variable force f directed along the axis and given by the formula $f = m_0 c^2/2\sqrt{(1 + x)}$. Show that the particle's velocity v is given by $v = x^{1/2}c/(1 + x)^{1/2}$. Putting $x = \sinh^2\theta$, if t is the time and $t = 0$ at O, prove that $ct = \theta + \sinh\theta\cosh\theta$.

67. A particle having rest mass m_0 is moving with speed $\tfrac{1}{2}c$ when it is subjected to a retarding force. When the particle's inertial mass is m, the magnitude of the retarding force is αm^2 (α is constant). Show that the time needed by the force to bring the particle to rest is $\pi c/6m_0\alpha$.

68. If T_{ij} is the energy–momentum tensor for an elastic fluid and V_i is its 4-velocity of flow, by verifying the equation $T_{ij}V_j = -c^2\mu_{00}V_i$ in a frame in which the fluid is momentarily at rest, prove it in any frame. Deduce the equations

$$g_\alpha = (\mu_{00}v_\alpha + \tau_{\alpha\beta}v_\beta/c^2)/(1 - v^2/c^2), \quad \mu = \mu_{00} + g_\alpha v_\alpha/c^2$$

Hence derive the following formulae for the elements of T_{ij}:

$$T_{\alpha\beta} = \mu_{00} V_\alpha V_\beta + \tau_{\alpha\beta} + \tau_{\alpha\gamma} V_\gamma V_\beta/c^2$$
$$T_{\alpha 4} = T_{4\alpha} = \mu_{00} V_\alpha V_4 + \tau_{\alpha\beta} V_\beta V_4/c^2$$
$$T_{44} = \mu_{00} V_4 V_4 - \tau_{\alpha\beta} V_\alpha V_\beta/c^2$$

69. A perfect fluid is streaming radially outwards across the surface of a sphere with radius R and centre O. If the motion is steady and there is no external force field, show that equation (21.20) leads to the equations

$$\frac{d}{dr}(r^3 v\lambda) + r^2\frac{dp}{dr} = 0, \quad r\frac{d\lambda}{dr} + 3\lambda = 0$$

where r is radial distance from O, p is the pressure, v is the speed of flow and $\lambda = (\mu_{00} + p/c^2)v/r(1 - v^2/c^2)$. If p vanishes over the sphere $r = R$ and $p = P$ at great distances, and if μ_{00} is constant outside the sphere, show that in this region

$$p = (P + c^2\mu_{00})\sqrt{(1 - v^2/c^2)} - c^2\mu_{00}$$
$$r^2 v(1 - v^2/c^2)^{-1/2} = R^2\sqrt{(P^2 + 2Pc^2\mu_{00})}/c\mu_{00}$$

70. A straight rod has cross-sectional area A and mass m per unit length. It lies along the x-axis of an inertial frame in a state of tension F. Show that the energy–momentum tensor has components which are the elements of a 4×4 diagonal matrix, with diagonal elements $(-F/A, 0, 0, -mc^2/A)$. Deduce that an observer moving along the x-axis with speed u, sees the inertial mass per unit length of the rod to be

$$\frac{m - Fu^2/c^4}{1 - u^2/c^2}$$

Deduce that F cannot exceed mc^2.

71. Assuming that the energy–momentum tensor T_{ij} is a tensor with respect to a general Lorentz transformation $\bar{x}_i = a_{ij}x_j + b_i$, write down the transformation equations for T_{ij} in the special case where $a_{4\alpha} = a_{\alpha 4} = 0$, $a_{44} = 1$. Deduce that $\mu, g_\alpha, g_{\alpha\beta}$ are 3-tensors with respect to a simple rotation of the frame $Ox_1x_2x_3$ without relative motion.

72. Relative to a frame S, a fluid has flow velocity $(u, 0, 0)$ at a certain point. In the frame S^0 relative to which the fluid is stationary at the point, the stress tensor has components $\tau_{\alpha\beta}^0$ and the fluid density is μ_{00}. Show that the energy–momentum tensor in the frame S has components

$$T_{11} = \frac{\tau_{11}^0 + \mu_{00}u^2}{1 - u^2/c^2}, \qquad T_{12} = (1 - u^2/c^2)^{-1/2}\tau_{12}^0,$$

$$T_{13} = (1 - u^2/c^2)^{-1/2}\tau_{13}^0, \qquad T_{14} = \frac{c^2\mu_{00} + \tau_{11}^0}{1 - u^2/c^2}\cdot\frac{iu}{c},$$

$$T_{21} = (1 - u^2/c^2)^{-1/2}\tau_{21}^0, \qquad T_{22} = \tau_{22}^0, \quad T_{23} = \tau_{23}^0,$$

$$T_{24} = (1 - u^2/c^2)^{-1/2} \tau_{12}^0 iu/c, \quad T_{31} = (1 - u^2/c^2)^{-1/2} \tau_{31}^0,$$

$$T_{32} = \tau_{32}^0, \quad T_{33} = \tau_{33}^0, \qquad\qquad T_{34} = (1 - u^2/c^2)^{-1/2} \tau_{13}^0 iu/c,$$

$$T_{44} = -\frac{c^2 \mu_{00} + \tau_{11}^0 u^2/c^2}{1 - u^2/c^2},$$

and deduce that

$$\tau_{11} = \tau_{11}^0, \quad \tau_{12} = (1 - u^2/c^2)^{-1/2} \tau_{12}^0, \quad \tau_{13} = (1 - u^2/c^2)^{-1/2} \tau_{13}^0,$$

$$\tau_{21} = (1 - u^2/c^2)^{1/2} \tau_{21}^0, \quad \tau_{22} = \tau_{22}^0, \quad \tau_{23} = \tau_{23}^0,$$

$$\tau_{31} = (1 - u^2/c^2)^{1/2} \tau_{31}^0, \quad \tau_{32} = \tau_{32}^0, \quad \tau_{33} = \tau_{33}^0,$$

$$\mu = (\mu_{00} + \tau_{11}^0 u^2/c^4)/(1 - u^2/c^2).$$

CHAPTER 4

Special Relativity Electrodynamics

24. 4-Current density

In this chapter we shall study the electromagnetic field due to a flow of charge which will be assumed known. Relative to an inertial frame S, let ρ be the charge density and $\mathbf{v}$ its velocity of flow. Then, if $\mathbf{j}$ is the current density,

$$\mathbf{j} = \rho \mathbf{v} \tag{24.1}$$

Assuming that charge can neither be created nor destroyed, the equation of continuity

$$\operatorname{div} \mathbf{j} + \frac{\partial \rho}{\partial t} = 0 \tag{24.2}$$

will be valid for the charge flow in S. This equation must be valid in every inertial frame and hence must be expressible in a form which is covariant with respect to orthogonal transformations in space–time. Introducing the coordinates x_i by equations (4.4) and employing equation (24.1), equation (24.2) is seen to be equivalent to

$$\frac{\partial}{\partial x_1}(\rho v_x) + \frac{\partial}{\partial x_2}(\rho v_y) + \frac{\partial}{\partial x_3}(\rho v_z) + \frac{\partial}{\partial x_4}(ic\rho) = 0 \tag{24.3}$$

This equation is covariant as required if $(\rho v_x,\ \rho v_y,\ \rho v_z,\ ic\rho)$ are the four components of a vector in space–time. For, if $\mathbf{J}$ is this vector, equation (24.3) can be written

$$J_{i,\,i} = 0 \tag{24.4}$$

and this is covariant with respect to orthogonal transformations. Now, by equation (15.6),

$$\mathbf{J} = (\rho \mathbf{v},\ ic\rho) = \rho(1 - v^2/c^2)^{1/2}\,\mathbf{V} \tag{24.5}$$

where $\mathbf{V}$ is the 4-velocity of flow and hence $\mathbf{J}$ is a vector if $\rho(1 - v^2/c^2)^{1/2}$ is an invariant. Denoting the invariant by ρ_0, we have

$$\rho = \frac{\rho_0}{(1 - v^2/c^2)^{1/2}} \tag{24.6}$$

73

It follows that $\rho = \rho_0$ if $v = 0$ and hence that ρ_0 is the charge density as measured from an inertial frame relative to which the charge being considered is instantaneously at rest. ρ_0 is called the *proper charge density*.

J is called the 4-*current density* and it is clear from equation (24.5) that

$$\mathbf{J} = \rho_0 \mathbf{V} = (\mathbf{j}, ic\rho) \qquad (24.7)$$

It is now clear that, when **J** has been specified throughout space–time, the charge flow is completely determined, for the space components of **J** fix the current density and the time component fixes the charge density. Hence, given **J**, the electromagnetic field must be calculable. The equations which form the basis for this calculation will be derived in the next two sections.

Let $d\omega_0$ be the volume of a small element of charge as measured from an inertial frame S_0 relative to which the charge is instantaneously at rest. The total charge within the element is $\rho_0 d\omega_0$. Due to the Fitzgerald contraction, the volume of this element as measured from S will be $d\omega$, where

$$d\omega = (1 - v^2/c^2)^{1/2} d\omega_0 \qquad (24.8)$$

The total charge within the element as measured from S is therefore

$$\rho d\omega = \rho(1 - v^2/c^2)^{1/2} d\omega_0 = \rho_0 d\omega_0 \qquad (24.9)$$

by equation (24.6). It follows that the electric charge on a body is invariant for all inertial observers.

25. 4-Vector potential

In classical theory, the equations determining the electromagnetic field due to a given charge flow are Maxwell's equations (3.1)–(3.4). To ensure covariance of the laws of mechanics with respect to Lorentz transformations, it proved necessary to modify classical Newtonian theory slightly. However, it will be shown that Maxwell's equations are covariant without any adjustment being necessary. Indeed, the Lorentz transformation equations were first noticed as the transformation equations which leave Maxwell's equations unaltered in form.

To prove this, it will be convenient to introduce the scalar and vector potentials, ϕ and **A** respectively, of the field. It is proved in textbooks devoted to the classical theory (Coulson and Boyd, 1979) that **A** satisfies the equations

$$\text{div } \mathbf{A} + \frac{1}{c^2}\frac{\partial \phi}{\partial t} = 0 \qquad (25.1)$$

$$\nabla^2 \mathbf{A} - \frac{1}{c^2}\frac{\partial^2 \mathbf{A}}{\partial t^2} = -\mu_0 \mathbf{j} \qquad (25.2)$$

and ϕ satisfies the equation

$$\nabla^2 \phi - \frac{1}{c^2}\frac{\partial^2 \phi}{\partial t^2} = -\rho/\varepsilon_0 \qquad (25.3)$$

where $c^2 = 1/\mu_0 \varepsilon_0$. We now define a 4-vector potential Ω in any inertial frame S by the equation

$$\Omega = (\mathbf{A}, i\phi/c) \tag{25.4}$$

It is easily verified that equations (25.2), (25.3) are together equivalent to the equation

$$\Box^2 \Omega = -\mu_0 \mathbf{J} \tag{25.5}$$

where the operator $\Box^2$ is defined by

$$\Box^2 = \frac{\partial^2}{\partial x_1^2} + \frac{\partial^2}{\partial x_2^2} + \frac{\partial^2}{\partial x_3^2} + \frac{\partial^2}{\partial x_4^2} \tag{25.6}$$

The space components of equation (25.5) yield equation (25.2) and the time component, equation (25.3). If Ω_i, J_i are the components of Ω and $\mathbf{J}$ respectively, equation (25.5) can be written

$$\Omega_{i,\,jj} = -\mu_0 J_i \tag{25.7}$$

in which form it is clearly covariant with respect to Lorentz transformations provided Ω is a vector. This confirms that equation (25.4) does, in fact, define a quantity with the transformation properties of a vector in space–time.

Next, it is necessary to show that equation (25.1) is also covariant with respect to orthogonal transformations in space–time. It is clearly equivalent to the equation

$$\operatorname{div} \Omega = \Omega_{i,\,i} = 0 \tag{25.8}$$

which is in the required form.

$\mathbf{J}$ being given, Ω is now determined by equations (25.7) and (25.8).

26. The field tensor

When $\mathbf{A}$ and ϕ are known in an inertial frame, the electric and magnetic intensities $\mathbf{E}$ and $\mathbf{B}$ respectively at any point in the electromagnetic field follow from the equations

$$\mathbf{E} = -\operatorname{grad} \phi - \frac{\partial \mathbf{A}}{\partial t} \tag{26.1}$$

$$\mathbf{B} = \operatorname{curl} \mathbf{A} \tag{26.2}$$

Making use of equations (4.4) and (25.4), these equations are easily shown to be equivalent to the set

$$\left. \begin{aligned} -\frac{i}{c} E_x &= \frac{\partial \Omega_4}{\partial x_1} - \frac{\partial \Omega_1}{\partial x_4} \\ -\frac{i}{c} E_y &= \frac{\partial \Omega_4}{\partial x_2} - \frac{\partial \Omega_2}{\partial x_4} \\ -\frac{i}{c} E_z &= \frac{\partial \Omega_4}{\partial x_3} - \frac{\partial \Omega_3}{\partial x_4} \end{aligned} \right\} \tag{26.3}$$

$$B_x = \frac{\partial \Omega_3}{\partial x_2} - \frac{\partial \Omega_2}{\partial x_3}$$
$$B_y = \frac{\partial \Omega_1}{\partial x_3} - \frac{\partial \Omega_3}{\partial x_1}$$
$$B_z = \frac{\partial \Omega_2}{\partial x_1} - \frac{\partial \Omega_1}{\partial x_2}$$

(26.4)

Equations (26.3) and (26.4) indicate that the six components of the vectors $-i\mathbf{E}/c$, $\mathbf{B}$ with respect to the rectangular Cartesian inertial frame S are the six distinct non-zero components in space–time of the skew-symmetric tensor $\Omega_{j,\ i} - \Omega_{i,\ j}$. We have proved, therefore, that equations (26.1), (26.2) are valid in all inertial frames if

$$(F_{ij}) = \begin{pmatrix} 0 & B_z & -B_y & -iE_x/c \\ -B_z & 0 & B_x & -iE_y/c \\ B_y & -B_x & 0 & -iE_z/c \\ iE_x/c & iE_y/c & iE_z/c & 0 \end{pmatrix}$$

(26.5)

is assumed to transform as a tensor with respect to orthogonal transformations in space–time. The equations (26.3) and (26.4) can then be summarized in the tensor equation

$$F_{ij} = \Omega_{j,\ i} - \Omega_{i,\ j}$$

(26.6)

F_{ij} is called the *electromagnetic field tensor*. The close relationship between the electric and magnetic aspects of an electromagnetic field is now revealed as being due to their both contributing as components to the field tensor which serves to unite them.

Consider now equations (3.2) and (3.3). Employing the field tensor defined by equation (26.5) and the current density given by equation (24.7), and recalling that $\mathbf{B} = \mu_0 \mathbf{H}$, $\mathbf{D} = \varepsilon_0 \mathbf{E}$, these equations are seen to be equivalent to

$$\frac{\partial F_{12}}{\partial x_2} + \frac{\partial F_{13}}{\partial x_3} + \frac{\partial F_{14}}{\partial x_4} = \mu_0 J_1$$
$$\frac{\partial F_{21}}{\partial x_1} + \frac{\partial F_{23}}{\partial x_3} + \frac{\partial F_{24}}{\partial x_4} = \mu_0 J_2$$
$$\frac{\partial F_{31}}{\partial x_1} + \frac{\partial F_{32}}{\partial x_2} + \frac{\partial F_{34}}{\partial x_4} = \mu_0 J_3$$
$$\frac{\partial F_{41}}{\partial x_1} + \frac{\partial F_{42}}{\partial x_2} + \frac{\partial F_{43}}{\partial x_3} = \mu_0 J_4$$

(26.7)

or, in short,

$$F_{ij,\ j} = \mu_0 J_i$$

(26.8)

an equation which is covariant with respect to Lorentz transformations.

Finally, consider equations (3.1) and (3.4). These can be written

$$\left.\begin{array}{l} \dfrac{\partial F_{34}}{\partial x_2} + \dfrac{\partial F_{42}}{\partial x_3} + \dfrac{\partial F_{23}}{\partial x_4} = 0 \\[2mm] \dfrac{\partial F_{41}}{\partial x_3} + \dfrac{\partial F_{13}}{\partial x_4} + \dfrac{\partial F_{34}}{\partial x_1} = 0 \\[2mm] \dfrac{\partial F_{12}}{\partial x_4} + \dfrac{\partial F_{24}}{\partial x_1} + \dfrac{\partial F_{41}}{\partial x_2} = 0 \\[2mm] \dfrac{\partial F_{23}}{\partial x_1} + \dfrac{\partial F_{31}}{\partial x_2} + \dfrac{\partial F_{12}}{\partial x_3} = 0 \end{array}\right\} \tag{26.9}$$

These equations are summarized thus:

$$F_{ij,\,k} + F_{jk,\,i} + F_{ki,\,j} = 0 \tag{26.10}$$

If any pair from i, j, k are equal, since F_{ij} is skew-symmetric, the left-hand member of this equation is identically zero and the equation is trivial. The four possible cases when i, j, k are distinct are the equations (26.9). Equation (26.10) is a tensor equation and is therefore also covariant with respect to Lorentz transformations.

To sum up, Maxwell's equations in 4-dimensional covariant form are:

$$\left.\begin{array}{l} F_{ij,\,j} = \mu_0 J_i \\[1mm] F_{ij,\,k} + F_{jk,\,i} + F_{ki,\,j} = 0 \end{array}\right\} \tag{26.11}$$

Given J_i at all points in space–time, these equations determine the field tensor F_{ij}. The solution can be found in terms of a vector potential Ω_i which satisfies the following equations:

$$\left.\begin{array}{l} \Omega_{i,\,i} = 0 \\[1mm] \Omega_{i,jj} = -\mu_0 J_i \end{array}\right\} \tag{26.12}$$

Ω_i being determined, F_{ij} follows from the equation

$$F_{ij} = \Omega_{j,\,i} - \Omega_{i,\,j} \tag{26.13}$$

27. Lorentz transformations of electric and magnetic vectors

Since F_{ij} is a tensor, relative to the special Lorentz transformation (5.1) its non-zero components transform thus:

$$\left.\begin{array}{l} \overline{F}_{23} = F_{23} \\[1mm] \overline{F}_{31} = F_{31}\cos\alpha + F_{34}\sin\alpha \\[1mm] \overline{F}_{12} = F_{12}\cos\alpha + F_{42}\sin\alpha \end{array}\right\} \tag{27.1}$$

$$\left.\begin{array}{l} \overline{F}_{14} = F_{14} \\[1mm] \overline{F}_{24} = -F_{21}\sin\alpha + F_{24}\cos\alpha \\[1mm] \overline{F}_{34} = -F_{31}\sin\alpha + F_{34}\cos\alpha \end{array}\right\} \tag{27.2}$$

Substituting for the components of F_{ij} from equation (26.5) and for $\sin \alpha$, $\cos \alpha$ from equations (5.7), the above equations (27.1) yield the special Lorentz transformation equations for **B**, viz.

$$\bar{B}_x = B_x, \quad \bar{B}_y = \beta(B_y + (u/c^2)E_z), \quad \bar{B}_z = \beta(B_z - (u/c^2)E_y) \qquad (27.3)$$

Similarly, equations (27.2) yield the transformation equations for **E**, viz.

$$\bar{E}_x = E_x, \quad \bar{E}_y = \beta(E_y - uB_z), \quad \bar{E}_z = \beta(E_z + uB_y) \qquad (27.4)$$

The inverse equations can be written down by exchanging 'barred' and 'unbarred' symbols and replacing u by $-u$.

As an example of the use to which these transformation formulae may be put, consider the electromagnetic field due to an infinitely long, uniformly charged wire lying at rest along the $\bar{x}$-axis of the inertial frame $\bar{S}$. If $\bar{q}$ is the charge per unit length, it is well known that the electric intensity is everywhere perpendicular to the wire and is of magnitude $\bar{q}/(2\pi\varepsilon_0\bar{r})$, where $\bar{r}$ is the perpendicular distance from the wire. Thus, at the point $(\bar{x}, \bar{y}, \bar{z})$, the components of $\bar{E}$ are given by

$$\bar{E}_x = 0, \quad \bar{E}_y = \frac{\bar{q}\bar{y}}{2\pi\varepsilon_0(\bar{y}^2 + \bar{z}^2)}, \quad \bar{E}_z = \frac{\bar{q}\bar{y}}{2\pi\varepsilon_0(\bar{y}^2 + \bar{z}^2)} \qquad (27.5)$$

The magnetic induction vanishes.

The electromagnetic field observed from the parallel inertial frame S (relative to which $\bar{S}$ has velocity $(u, 0, 0)$) is given by the inverses of equations (27.3) and (27.4) to have components

$$\begin{aligned}
B_x = 0, \quad B_y &= -\frac{\beta u \bar{q} z}{2\pi\varepsilon_0 c^2 (y^2 + z^2)}, \quad B_z = \frac{\beta u \bar{q} y}{2\pi\varepsilon_0 c^2 (y^2 + z^2)} \\[2mm]
E_x = 0, \quad E_y &= \frac{\beta \bar{q} y}{2\pi\varepsilon_0 (y^2 + z^2)}, \quad E_z = \frac{\beta \bar{q} z}{2\pi\varepsilon_0 (y^2 + z^2)}
\end{aligned} \right\} \qquad (27.6)$$

at the point (x, y, z) (having used the transformation equations $y = \bar{y}, z = \bar{z}$). A segment of the wire having unit length in $\bar{S}$ will appear in S to have length $\sqrt{(1 - u^2/c^2)}$; however, the charge on the segment must be the same in both frames, viz. $\bar{q}$. It follows that the charge per unit length as observed in S is $q = \beta\bar{q}$. Thus, the charge which flows past a fixed point on the x-axis of S in unit time will be $\beta u\bar{q} = i$; i therefore measures the current flowing along this axis. Since $c^2 = 1/\mu_0\varepsilon_0$, equations (27.6) can now be written

$$\begin{aligned}
B_x = 0, \quad B_y &= -\frac{\mu_0 i z}{2\pi(y^2 + z^2)}, \quad B_z = \frac{\mu_0 i y}{2\pi(y^2 + z^2)} \\[2mm]
E_x = 0, \quad E_y &= \frac{qy}{2\pi\varepsilon_0(y^2 + z^2)}, \quad E_z = \frac{qz}{2\pi\varepsilon_0(y^2 + z^2)}
\end{aligned} \right\} \qquad (27.7)$$

These equations imply that the magnetic induction is of magnitude $\mu_0 i/2\pi r$ and that the **B**-lines are circles with centres on Ox and planes parallel to Oyz. This result for a long straight current is a well-known one in the classical theory. The electric intensity is of magnitude $q/2\pi\varepsilon_0 r$ and is directed radially from the wire; however, in the case of a current due to the flow of negatively charged electrons in a stationary wire, this field is cancelled by the contrary field due to the positive charges on the atomic nuclei.

28. The Lorentz force

We shall now calculate the force exerted upon a point charge e in motion in an electromagnetic field.

At any instant, we can choose an inertial frame relative to which the point charge is instantaneously at rest. Let $\mathbf{E}_0$ be the electric intensity at the point charge relative to this frame. Then, by the physical definition of electric intensity as the force exerted upon unit stationary charge, the force exerted upon e will be $e\mathbf{E}_0$. It follows from equation (18.2) that the 4-force acting upon the charge in this frame is given by

$$\mathbf{F} = (e\mathbf{E}_0, 0) \tag{28.1}$$

The 4-velocity of the charge in this frame is also given by

$$\mathbf{V} = (0, ic) \tag{28.2}$$

and hence, by equation (26.5),

$$eF_{ij}V_j = e(E_{x0}, E_{y0}, E_{z0}, 0) = (e\mathbf{E}_0, 0) \tag{28.3}$$

It has accordingly been shown that, in an inertial frame relative to which the charge is instantaneously stationary,

$$F_i = eF_{ij}V_j \tag{28.4}$$

But this is an equation between tensors and is therefore true for all inertial frames.

Substituting in equation (28.4) for the components F_i, F_{ij}, V_j from equations (18.2), (26.5) and (15.6) respectively, the following equations are obtained:

$$\left.\begin{array}{l} f_x = e(B_z v_y - B_y v_z + E_x) \\ f_y = e(B_x v_z - B_z v_x + E_y) \\ f_z = e(B_y v_x - B_x v_y + E_z) \end{array}\right\} \tag{28.5}$$

These equations are equivalent to the 3-vector equation

$$\mathbf{f} = e(\mathbf{E} + \mathbf{v} \times \mathbf{B}) \tag{28.6}$$

$\mathbf{f}$ is called the *Lorentz force* acting upon the charged particle.

29. The energy–momentum tensor for an electromagnetic field

Suppose that a charge distribution is specified by a 4-current density vector $\mathbf{J}$. If $d\omega_0$ is the proper volume of any small element of the distribution and ρ_0 is the proper density of the charge, the charge within the element will be $\rho_0 d\omega_0$. It follows from equation (28.4) that the 4-force exerted upon the element by the electromagnetic field is given by

$$F_i = \rho_0 F_{ij} V_j d\omega_0 \qquad (29.1)$$

$\mathbf{V}$ being the 4-velocity of flow for the element. Employing equation (24.7), this last equation can be written

$$F_i = F_{ij} J_j d\omega_0 \qquad (29.2)$$

and it follows from the definition given in section 21 that the 4-force density for the electromagnetic field is given by

$$D_i = F_{ij} J_j \qquad (29.3)$$

Substituting for J_j from the first of equations (26.11), we can express D_i in terms of the field tensor thus:

$$D_i = \frac{1}{\mu_0} F_{ij} F_{jk,k} \qquad (29.4)$$

We will now prove that the right-hand member of this equation is, apart from sign, the divergence of a certain symmetric tensor S_{ij} given by the equation

$$\mu_0 S_{ij} = F_{ik} F_{jk} - \tfrac{1}{4} \delta_{ij} F_{kl} F_{kl} \qquad (29.5)$$

and called the *energy–momentum tensor* of the electromagnetic field.

Taking the divergence of S_{ij}, we have

$$\mu_0 S_{ij,j} = F_{ik,j} F_{jk} + F_{ik} F_{jk,j} - \tfrac{1}{2} \delta_{ij} F_{kl} F_{kl,j} \qquad (29.6)$$

Now

$$F_{ik,j} F_{jk} = F_{ij,k} F_{kj} = F_{ji,k} F_{jk} \qquad (29.7)$$

since F_{ij} is skew-symmetric. Thus

$$F_{ik,j} F_{jk} = \tfrac{1}{2} (F_{ik,j} + F_{ji,k}) F_{jk} \qquad (29.8)$$

Also

$$\delta_{ij} F_{kl} F_{kl,j} = F_{kl} F_{kl,i} = -F_{jk} F_{kj,i} \qquad (29.9)$$

and it follows from these results that the first and last terms of the right-hand member of equation (29.6) can be combined to yield

$$\tfrac{1}{2} (F_{ik,j} + F_{ji,k} + F_{kj,i}) F_{jk} \qquad (29.10)$$

and this is zero by the second of equations (26.11).

Hence

$$S_{ij,j} = \frac{1}{\mu_0} F_{ik} F_{jk,j} = -\frac{1}{\mu_0} F_{ik} F_{kj,j} = -D_i \qquad (29.11)$$

Substituting for the components of the field tensor from equation (26.5), the components of S_{ij} are calculable from equation (29.5) as follows: If α, β take any of the values 1, 2, 3, then writing E_1 for E_x, E_2 for E_y, etc.,

$$S_{\alpha\beta} = -(\varepsilon_0 E_\alpha E_\beta + \mu_0 H_\alpha H_\beta), \qquad \alpha \neq \beta \tag{29.12}$$

If $i = j = 1$,

$$S_{11} = -(\mu_0 H_1^2 + \varepsilon_0 E_1^2) + \tfrac{1}{2}(\mu_0 H^2 + \varepsilon_0 E^2) \tag{29.13}$$

S_{22}, S_{33} may be expressed similarly and therefore, in general, if $\alpha, \beta, = 1, 2, 3$,

$$S_{\alpha\beta} = -(\varepsilon_0 E_\alpha E_\beta + \mu_0 H_\alpha H_\beta) + \tfrac{1}{2}\delta_{\alpha\beta}(\varepsilon_0 E^2 + \mu_0 H^2) \tag{29.14}$$

Apart from sign, this is *Maxwell's stress tensor* t_{ij}. t_{ij} is only a tensor with respect to rectangular frames stationary in the inertial frame being employed.

Also, if $\alpha = 1, 2, 3$,

$$S_{\alpha 4} = S_{4\alpha} = \frac{i}{c}(E_2 H_3 - E_3 H_2, E_3 H_1 - E_1 H_3, E_1 H_2 - E_2 H_1)$$

$$= \frac{i}{c}\mathbf{E} \times \mathbf{H} = \frac{i}{c}\mathbf{S} \tag{29.15}$$

where $\mathbf{S}$ is *Poynting's vector*.

Finally,

$$S_{44} = -\tfrac{1}{2}(\varepsilon_0 E^2 + \mu_0 H^2) = -U \tag{29.16}$$

where U is the energy density in the electromagnetic field.

These results may be summarized conveniently by exhibiting the components of S_{ij} in a matrix thus:

$$(S_{ij}) = -\left(\frac{t_{ij}}{\mathbf{S}/ic} \,\middle|\, \frac{\mathbf{S}/ic}{U}\right) \tag{29.17}$$

We can now write down the equation of motion for a charge cloud moving under the action of the electromagnetic field it generates. If T_{ij} is the kinetic energy–momentum tensor for the cloud, equations (21.15) and (29.11) show that the equation of motion can be written

$$T_{ij, j} = -S_{ij, j} \tag{29.18}$$

or

$$(T_{ij} + S_{ij}), j = 0 \tag{29.19}$$

i.e. the divergence of the total energy–momentum tensor vanishes. If the charged particles forming the cloud do not interact except via the electromagnetic field, i.e. the cloud is incoherent, T_{ij} is given by equation (21.16). If, however, the particles constitute an ionized fluid, equations (22.19) or (22.21) must be used to calculate T_{ij}.

It was shown in section 21 that $T_{\alpha 4}/ic$ equals the density of the x_α-component of the linear momentum of a system. Since $S_{\alpha 4}/ic = S_\alpha/c^2$, the density of the linear momentum of an electromagnetic field is $\mathbf{g} = \mathbf{S}/c^2$, where $\mathbf{S}$ is Poynting's vector. Alternatively, as explained in section 21, $\mathbf{g}$ can be interpreted as the current density vector for the inertial mass flow and thus, $c^2\mathbf{g} = \mathbf{S}$ gives the rate of energy flow across unit area placed perpendicular to the direction of this flow; this is the usual significance attached to Poynting's vector.

According to the theory in section 21, $-S_{44}/c^2$ should equal the density of inertial mass for the field. We have found that $-S_{44}/c^2 = U/c^2$ and, since U is the energy density, our result is in conformity with expectations.

The results which have been obtained may be summarized as follows: If momentum of density $\mathbf{S}/c^2$ and energy of density U are ascribed to the electromagnetic field, equation (29.19) shows that the net momentum and energy of the field and charge will be conserved.

Exercises 4

1. Write down the special Lorentz transformation equations for $\mathbf{J}$ and deduce the transformation equations for $\mathbf{j}$, ρ, viz.

$$\bar{j}_x = (1 - u^2/c^2)^{-1/2}(j_x - \rho u), \quad \bar{j}_y = j_y$$
$$\bar{\rho} = (1 - u^2/c^2)^{-1/2}(\rho - j_x u/c^2), \quad \bar{j}_z = j_z$$

2. Deduce from the Maxwell equation $F_{ij,j} = \mu_0 J_i$ that div $\mathbf{J} = 0$.

3. Verify that the field tensor defined in terms of the 4-potential Ω_i by equation (26.13) satisfies Maxwell's equations (26.11) provided Ω_i satisfies the equations (26.12).

4. (i) Prove that

$$F_{ij}F_{ij} = 2\mu_0(\mu_0 H^2 - \varepsilon_0 E^2)$$

and deduce that $\mu_0 H^2 - \varepsilon_0 E^2$ is invariant with respect to Lorentz transformations.

(ii) Prove that

$$e_{ijkl}F_{ij}F_{kl} = -8i\mathbf{E} \cdot \mathbf{B}/c$$

and deduce that $\mathbf{E} \cdot \mathbf{B}$ is an invariant density with respect to Lorentz transformations.

5. If U is the energy density and $\mathbf{S}$ is the Poynting vector for an electromagnetic field, prove that $U^2 - \mathbf{S}^2$ is invariant.

6. An observer O at rest in an inertial frame O$xyzt$ finds himself to be in an electric field $\mathbf{E} = (0, E, 0)$, with no magnetic field. Show that an observer $\overline{\text{O}}$ moving according to O with uniform velocity $\mathbf{V}$ at right angles to $\mathbf{E}$, finds electric and magnetic fields $\overline{\mathbf{E}}$, $\overline{\mathbf{B}}$ connected by the relation

$$c^2\overline{\mathbf{B}} + \mathbf{V} \times \overline{\mathbf{E}} = 0$$

7. If S_{ij} is the energy–momentum tensor for an electromagnetic field, prove that its trace, viz. S_{ii}, is zero.

8. A plane monochromatic electromagnetic wave is being propagated in a direction parallel to the x-axis in the inertial frame S. Its electric and magnetic field components are given by

$$\mathbf{E} = [0, a\sin\omega(t - x/c), 0]$$

$$\mathbf{B} = [0, 0, \frac{a}{c}\sin\omega(t - x/c)]$$

Show that, when observed from the inertial frame $\bar{S}$, it appears as the plane monochromatic wave

$$\bar{\mathbf{E}} = [0, \lambda a\sin\lambda\omega(\bar{t} - \bar{x}/c), 0]$$

$$\bar{\mathbf{B}} = [0, 0, \lambda\frac{a}{c}\sin\lambda\omega(\bar{t} - \bar{x}/c)]$$

where

$$\lambda = \sqrt{\left(\frac{1 - u/c}{1 + u/c}\right)}$$

u being the velocity of $\bar{S}$ relative to S (i.e. both the amplitude and frequency are reduced by a factor λ. The reduction in frequency is the *Doppler effect*.)

9. Show that the Hamiltonian for the motion of a particle with charge e and mass m in an electromagnetic field $(\mathbf{A}, \phi)$ is

$$H = c\left[\left(\mathbf{p} - e\mathbf{A}\right)^2 + m^2c^2\right]^{1/2} + e\phi$$

(*Hint*: Show that Hamilton's equations yield the equation

$$\frac{d}{dt}(m\mathbf{v}) = e(\mathbf{E} + \mathbf{v} \times \mathbf{B}).)$$

10. Verify that, in a region devoid of charge, equations (26.12) are satisfied by

$$\Omega_i = A_i e^{ik_p x_p}$$

provided A_i, k_p are constants such that

$$A_i k_i = 0, \quad k_p k_p = 0$$

By considering the 4-vector property of Ω_i, deduce that A_i must transform as a 4-vector under Lorentz transformations. Deduce also that $k_p x_p$ is a scalar under such transformations and hence that k_p is a 4-vector.

A plane electromagnetic wave, whose direction of propagation is parallel to the plane Oxy and makes an angle α with Ox, is given by

$$\Omega_i = A_i e^{2\pi i v(x\cos\alpha + y\sin\alpha - ct)/c}$$

where v is the frequency. The same wave observed from a parallel frame $\overline{Oxyz}$ moving with velocity u along Ox, has frequency $\bar{v}$ and direction of propagation making an angle $\bar{\alpha}$ with $\overline{Ox}$. By writing down the transformation equations for the vector k_p, prove that

$$\bar{v} = \frac{1 - \dfrac{u}{c}\cos\alpha}{(1 - u^2/c^2)^{1/2}}v, \quad \cos\bar{\alpha} = \frac{\cos\alpha - \dfrac{u}{c}}{1 - \dfrac{u}{c}\cos\alpha}$$

11. $Oxyz$ is an inertial frame S. A particle having rest mass m_0 and electric charge q moves in the xy-plane under the action of a uniform magnetic field B directed along the z-axis. Show that the particle's speed v is constant and that, with a suitable choice of coordinates, its trajectory is the circle

$$x = R\sin\omega t, \quad y = R\cos\omega t$$

where

$$\omega = qB/\beta m_0, \quad R = v/\omega, \quad \beta = (1 - v^2/c^2)^{-1/2}$$

$\bar{S}$ is an inertial frame $\bar{O}\bar{x}\bar{y}\bar{z}$ parallel to S and $\bar{O}$ moves along Ox with speed u. Calculate u and B so that uniform fields $\mathbf{E} = (0, E_0, 0)$, $\mathbf{B} = (0, 0, B_0)$ are observed in $\bar{S}$. Hence describe the motion of a charged particle released in these fields and show that its average velocity is E_0/B_0 along the $\bar{x}$-axis.

12. The frame $\bar{S}$ is parallel to the frame S and is moving along the x-axis with speed u. In the frame S, there is a uniform electric field $(0, E, 0)$ and a uniform magnetic field $(0, 0, B)$. Show that it is possible to choose the value of u so that the field in the frame $\bar{S}$ is entirely magnetic and that its magnitude is then $\sqrt{(B^2 - E^2/c^2)}$. What is the direction of this field? (Ans. Parallel to $\bar{z}$-axis.)

13. A charge q has rest mass m_0 and is moving in the positive sense along a negative x-axis with speed u, when it enters a magnetic field, having components $B_x = B_y = 0$, $B_z = B$ (constant), confined to the region $0 \leqslant x \leqslant a$. There is no field in the regions $x < 0$, $x > a$. Explain why the inertial mass of the charge remains constant during its motion through the field and show that its path is the circle

$$x^2 + y^2 + 2ky = 0, \quad z = 0$$

where $k = m_0 u/qB(1 - u^2/c^2)^{1/2}$. What is the condition that the charge will be turned back by the field? (Ans. $k < a$.)

14. A plane electromagnetic wave of frequency f is being propagated in a direction making an angle θ with the x-axis. Its electric and magnetic field components are given by

$$\mathbf{E} = (-AX\sin\theta, AX\cos\theta, 0)$$
$$\mathbf{B} = (0, 0, AX/c)$$

where

$$X = \sin 2\pi f\left(t - \frac{x\cos\theta + y\sin\theta}{c}\right)$$

Show that, when observed from a frame $\bar{S}$ which is parallel to $Oxyz$ and moves with a velocity $(c\cos\theta, 0, 0)$ relative to $Oxyz$, the wave has components

$$\bar{\mathbf{E}} = (-A\bar{X}\sin\theta, 0, 0)$$
$$\bar{\mathbf{B}} = (0, 0, A\bar{X}\sin\theta/c)$$

where

$$\bar{X} = \sin[2\pi f \sin\theta(\bar{t} - \bar{y}/c)]$$

and $(\bar{x}, \bar{y}, \bar{z}, \bar{t}\,)$ are space–time coordinates in $\bar{S}$. What is the direction of propagation in $\bar{S}$ and what is the observed frequency? (Ans. Parallel to $\bar{y}$-axis at frequency $f\sin\theta$.)

15. V_i is the 4-velocity of flow of a conducting medium and J_i is the 4-current density of a charge flow in the medium. Ohm's law is valid for the medium, σ being its conductivity. Prove that

$$J_i + \frac{1}{c^2}V_i J_j V_j = \sigma F_{ij} V_j$$

(*Hint*: Verify this equation in a frame for which the medium is at rest using Ohm's law $\mathbf{j} = \sigma\mathbf{E}$.)

16. A uniform magnetic field of induction B is directed along the z-axis of an inertial frame. Show that the energy–momentum tensor for the field has components which are the elements of a 4×4 diagonal matrix, with diagonal elements $B^2/2\mu_0(1, 1, -1, -1)$.

17. A point charge e moves along the z-axis of an inertial frame S with constant velocity $\mathbf{v}$. Calculate the electromagnetic field in a parallel inertial frame whose origin moves with the charge and deduce the field in S. Hence show that, at the instant $t = 0$, when the charge passes through the origin O of S, the electric field is directed radially from O and its magnitude at the point having spherical polar coordinates (r, θ, ϕ) is given by

$$E = \frac{e}{4\pi\varepsilon_0 r^2}(1 - v^2/c^2)(1 - v^2\sin^2\theta/c^2)^{-3/2}$$

Show, also, that the magnetic field in S at this instant is given by $\mathbf{B} = (\mathbf{v} \times \mathbf{E})/c^2$.

CHAPTER 5

General Tensor Calculus. Riemannian Space

30. Generalized N-dimensional spaces

In Chapter 2 the theory of tensors was developed in an N-dimensional Euclidean space on the understanding that the coordinate frame being employed was always rectangular Cartesian. If x_i, $x_i + dx_i$ are the coordinates of two neighbouring points relative to such a frame, the 'distance' ds between them is given by the equation

$$ds^2 = dx_i dx_i \qquad (30.1)$$

If $\bar{x}_i$, $\bar{x}_i + d\bar{x}_i$ are the coordinates of the same points with respect to another rectangular Cartesian frame, then

$$ds^2 = d\bar{x}_i d\bar{x}_i \qquad (30.2)$$

and it follows that the expression $dx_i dx_i$ is invariant with respect to a transformation of coordinates from one rectangular Cartesian frame to another. Such a transformation was termed orthogonal.

Now, even in $\mathscr{E}_3$, it is very often convenient to employ a coordinate frame which is not Cartesian. For example, spherical polar coordinates (r, θ, ϕ) are frequently introduced, these being related to rectangular Cartesian coordinates (x, y, z) by the equations

$$x = r \sin\theta \cos\phi, \qquad y = r \sin\theta \sin\phi, \qquad z = r \cos\theta \qquad (30.3)$$

In such coordinates, the expression for ds^2 will be found to be

$$ds^2 = dx^2 + dy^2 + dz^2$$
$$= dr^2 + r^2 d\theta^2 + r^2 \sin^2\theta \, d\phi^2 \qquad (30.4)$$

and this is no longer of the simple form of equation (30.1). The coordinate transformation (30.3) is accordingly not orthogonal. In fact, it is not even linear, as was the most general coordinate transformation (8.1) considered in Chapter 2.

The spherical polar coordinate system is an example of a *curvilinear coordinate frame* in $\mathscr{E}_3$. Let (u, v, w) be quantities related to rectangular Cartesian coordinates (x, y, z) by equations

$$u = u(x, y, z), \quad v = v(x, y, z), \quad w = w(x, y, z) \qquad (30.5)$$

such that, to each point there corresponds a unique triad of values of (u, v, w) and to each such triad there corresponds a unique point. Then a set of values of (u, v, w) will serve to identify a point in $\mathscr{E}_3$ and (u, v, w) can be employed as coordinates. Such generalized coordinates are called *curvilinear coordinates*.

The equation

$$u(x, y, z) = u_0 \tag{30.6}$$

where u_0 is some constant, defines a surface in $\mathscr{E}_3$ over which u takes the constant value u_0. Similarly, the equations

$$v = v_0, \quad w = w_0 \tag{30.7}$$

define a pair of surfaces on which v takes the value v_0 and w the value w_0 respectively. These three surfaces will all pass through the point P_0 having coordinates (u_0, v_0, w_0) as shown in Fig. 6. They are called the *coordinate surfaces* through P_0. The surfaces $v = v_0, w = w_0$ will intersect in a curve $P_0 U$ along which v and w will be constant in value and only u will vary. $P_0 U$ is a *coordinate line* through P_0. Altogether, three coordinate lines pass through P_0. The equations u = constant, v = constant, w = constant define three families of coordinate surfaces corresponding to the three families of planes parallel to the coordinate planes $x = 0, y = 0, z = 0$ of a rectangular Cartesian frame. Pairs of these surfaces intersect in coordinate lines which correspond to the parallels to the coordinate axes in a Cartesian frame.

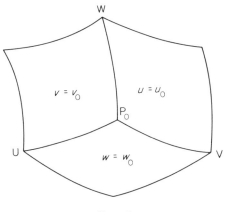

FIG. 6

Solving equations (30.5) for (x, y, z) in terms of (u, v, w), we obtain the inverse transformation

$$x = x(u, v, w), \quad y = y(u, v, w), \quad z = z(u, v, w) \tag{30.8}$$

Let (x, y, z), $(x + dx, y + dy, z + dz)$ be the rectangular Cartesian coordinates of two neighbouring points and let (u, v, w), $(u + du, v + dv, w + dw)$ be their respective curvilinear coordinates. Differentiating equations (30.8), we obtain

$$dx = \frac{\partial x}{\partial u}du + \frac{\partial x}{\partial v}dv + \frac{\partial x}{\partial w}dw, \quad \text{etc.} \quad (30.9)$$

Thus, if ds is the distance between these points,

$$ds^2 = dx^2 + dy^2 + dz^2$$
$$= Adu^2 + Bdv^2 + Cdw^2 + 2Fdvdw + 2Gdwdu + 2Hdudv \quad (30.10)$$

giving the appropriate expression for ds^2 in curvilinear coordinates. It will be noted that the coefficients A, B, etc., are, in general, functions of (u, v, w).

If, therefore, curvilinear coordinate frames are to be permitted, the theory of tensors developed in Chapter 2 must be modified to make it independent of the special orthogonal transformations for which ds^2 is always expressible in the simple form of equation (30.1). The necessary modifications will be described in the later sections of this chapter. However, these modifications prove to be of such a nature that the amended theory makes no appeal to the special metrical properties of Euclidean space, i.e. the theory proves to be applicable in more general spaces for which Euclidean space is a particular case. This we shall now explain further.

Let $(x^1, x^2, \ldots, x^N)$ be curvilinear coordinates in $\mathscr{E}_N$.* Then, by analogy with equation (30.10), if ds is the distance between two neighbouring points, it can be shown that

$$ds^2 = g_{ij}dx^i dx^j \quad (30.11)$$

where the coefficients g_{ij} of the quadratic form in the x^j will, in general, be functions of these coordinates. Since the space is Euclidean, it is possible to transform from the curvilinear coordinates x^j to Cartesian coordinates y^i so that

$$ds^2 = dy^i dy^i \quad (30.12)$$

Clearly, the reduction of ds^2 to this simple form is only possible because the functions g_{ij} satisfy certain conditions. Conversely, the satisfaction of these conditions by the g_{ij} will guarantee that coordinates y^i exist for which ds^2 takes the simple form (30.12) and hence that the space is Euclidean. However, in extending the theory of tensors to be applicable to curvilinear coordinate frames, we shall, at a certain stage, make use of the fact that ds^2 is expressible in the form (30.11), but no use will be made of the conditions satisfied by the coefficients g_{ij} which are a consequence of the space being Euclidean. It follows that the extended theory will be applicable in a hypothetical N-dimensional space for which the 'distance' ds between neighbouring points x^i, $x^i + dx^i$ is given by an equation (30.11) in which the g_{ij} are *arbitrary functions* of the x^i.† Such a space is

---

* The coordinates are here distinguished by superscripts instead of subscripts for a reason which will be given later.
† Except that partial derivatives of the g_{ij} will be assumed to exist and to be continuous to any order required by the theory.

said to be *Riemannian* and will be denoted by $\mathcal{R}_N$. $\mathcal{E}_N$ is a particular $\mathcal{R}_N$ for which the g_{ij} satisfy certain conditions. The right-hand member of equation (30.11) is termed the *metric* of the Riemannian space.

The surface of the Earth provides an example of an $\mathcal{R}_2$. If θ is the co-latitude and ϕ is the longitude of any point on the Earth's surface, the distance ds between the points (θ, ϕ), $(\theta + d\theta, \phi + d\phi)$ is given by

$$\mathrm{d}s^2 = R^2 (\mathrm{d}\theta^2 + \sin^2\theta \mathrm{d}\phi^2) \tag{30.13}$$

where R is the earth's radius. For this space and coordinate frame, the g_{ij} take the form

$$g_{11} = R^2, \quad g_{12} = g_{21} = 0, \quad g_{22} = R^2 \sin^2\theta \tag{30.14}$$

It is not possible to define other coordinates (x, y) in terms of which

$$\mathrm{d}s^2 = \mathrm{d}x^2 + \mathrm{d}y^2 \tag{30.15}$$

over the whole surface, i.e. this $\mathcal{R}_2$ is not Euclidean. However, the surfaces of a right circular cylinder and cone are Euclidean; the proof is left as an exercise for the reader.

It will be proved in Chapter 6 that, in the presence of a gravitational field, space–time ceases to be Euclidean in Minkowski's sense and becomes an $\mathcal{R}_4$. This is our chief reason for considering such spaces. However, we can generalize the concept of the space in which our tensors are to be defined yet further. Until section 37 is reached, we shall make no further reference to the metric. This implies that the theory of tensors, as developed thus far, is applicable in a very general N-dimensional space in which it is assumed it is possible to set up a coordinate frame but which is not assumed to possess a metric. In such a hypothetical space, the distance between two points is not even defined. It will be referred to as $\mathcal{S}_N$. $\mathcal{R}_N$ is a particular $\mathcal{S}_N$ for which a metric is specified.

31. Contravariant and covariant tensors

Let x^i be the coordinates of a point P in $\mathcal{S}_N$ relative to a coordinate frame which is specified in some manner which does not concern us here. Let $\bar{x}^i$ be the coordinates of the same point with respect to another reference frame and let these two systems of coordinates be related by equations

$$\bar{x}^i = \bar{x}^i(x^1, x^2, \ldots, x^N) \tag{31.1}$$

Consider the neighbouring point P' having coordinates $x^i + \mathrm{d}x^i$ in the first frame. Its coordinates in the second frame will be $\bar{x}^i + \mathrm{d}\bar{x}^i$, where

$$\mathrm{d}\bar{x}^i = \frac{\partial \bar{x}^i}{\partial x^j}\mathrm{d}x^j \tag{31.2}$$

and summation with respect to the index j is understood. The N quantities $\mathrm{d}x^i$ are taken to be the components of the *displacement vector* PP' referred to the first frame. The components of this vector referred to the second frame are, correspondingly, the $d\bar{x}^i$ and these are related to the components in the first frame by the transformation equation (31.2). Such a displacement vector is taken to be the prototype for all *contravariant vectors*.

Thus, A^i are said to be the components of a contravariant vector located at the point x^i, if the components of the vector in the 'barred' frame are given by the equation

$$\bar{A}^i = \frac{\partial \bar{x}^i}{\partial x^j} A^j \tag{31.3}$$

It is important to observe that, whereas in Chapter 2 the coefficients a_{ij} occurring in the transformation equation (10.2) were not functions of the Cartesian coordinates x_i so that the vector A was not, necessarily, located at a definite point in $\mathscr{E}_N$, the coefficients $\partial \bar{x}^i / \partial x^j$ in the corresponding equation (31.3) are functions of the x^i and the precise location of the vector A^i must be known before its transformation equations are determinate. This can be expressed by saying that there are no *free vectors* in $\mathscr{S}_N$.

The form of the transformation equation (31.3) should be studied carefully. It will be observed that the dummy index j occurs once as a superscript and once as a subscript (i.e. in the denominator of the partial derivative). Dummy indices will invariably occupy such positions in all expressions with which we shall be concerned. Again, the free index i occurs as a superscript on both sides of the equation. This rule will be followed in all later developments, i.e. a free index will always occur in the same position (upper or lower) in each term of an equation. Finally, it will assist the reader to memorize this transformation if he notes that the free index is associated with the 'barred' symbol on both sides of the equation.

A contravariant vector A^i may be defined at one point of $\mathscr{S}_N$ only. However, if it is defined at every point of a certain region, so that the A^i are functions of the x^i, a *contravariant vector field* is said to exist in the region.

If V is a quantity which is unaltered in value when the reference frame is changed, it is said to be a *scalar* or an *invariant* in $\mathscr{S}_N$. Its transformation equation is simply

$$\bar{V} = V \tag{31.4}$$

Since this equation involves no coefficients dependent upon the x^i, the possibility that V may be a free invariant exists. However, V is more often associated with a specific point in $\mathscr{S}_N$ and may be defined at all points of a region of $\mathscr{S}_N$, in which case it defines an *invariant field*. In the latter case

$$V = V(x^1, x^2, \ldots, x^N) \tag{31.5}$$

$\bar{V}$ will then, in general, be a quite distinct function of the $\bar{x}^i$. If, however, in this function we substitute for the $\bar{x}^i$ in terms of the x^i from equation (31.1), by equation (31.4) the right-hand member of equation (31.5) must result. Thus

$$\bar{V}(\bar{x}^1, \bar{x}^2, \ldots, \bar{x}^N) = V(x^1, x^2, \ldots, x^N) \tag{31.6}$$

V being an invariant field, consider the N derivatives $\partial V/\partial x^i$. In the $\bar{x}^i$-frame, the corresponding quantities are $\partial \bar{V}/\partial \bar{x}^i$ and we have

$$\frac{\partial \bar{V}}{\partial \bar{x}^i} = \frac{\partial \bar{V}}{\partial x^j}\frac{\partial x^j}{\partial \bar{x}^i} = \frac{\partial x^j}{\partial \bar{x}^i}\frac{\partial V}{\partial x^j} \tag{31.7}$$

since, by equation (31.6), when $\bar{V}$ is expressed as a function of the x^i it reduces to V. As in Chapter 2, the $\partial V/\partial x^i$ are taken to be the components of a vector called the *gradient* of V and denoted by grad V. However, its transformation law (31.7) is not the same as that for a contravariant vector, viz. (31.3), and it is taken to be the prototype for another species of vectors called *covariant vectors*.

Thus, B_i is a covariant vector if

$$\bar{B}_i = \frac{\partial x^j}{\partial \bar{x}^i} B_j \tag{31.8}$$

Covariant vectors will be distinguished from contravariant vectors by writing their components with subscripts instead of superscripts. This notation is appropriate, for $\partial V/\partial x^i$ is a covariant vector and the index i occurs in the denominator of this partial derivative. The vector $\mathrm{d}x^i$, on the other hand, has been shown to be contravariant in its transformation properties and this is correctly indicated by the upper position of the index. This is the reason for denoting the coordinates by x^i instead of x_i, although it must be clearly understood that the x^i alone are not the components of a vector at all.

The reader should check that the three rules formulated above in relation to the transformation equation (31.3), apply equally to the equation (31.8).

The generalization from vectors to tensors now proceeds along the same lines as in section 10. Thus, if A^i, B^j are two contravariant vectors, the N^2 quantities $A^i B^j$ are taken as the components of a contravariant tensor of the second rank. Its transformation equation is found to be

$$\bar{A}^i \bar{B}^j = \frac{\partial \bar{x}^i}{\partial x^k}\frac{\partial \bar{x}^j}{\partial x^l} A^k B^l \tag{31.9}$$

Any set of N^2 quantities C^{ij} transforming in this way is a *contravariant tensor*.

Again, if A^i, B_j are vectors, the first contravariant and the second covariant, then the N^2 quantities $A^i B_j$ transform thus:

$$\bar{A}^i \bar{B}_j = \frac{\partial \bar{x}^i}{\partial x^k}\frac{\partial x^l}{\partial \bar{x}^j} A^k B_l \tag{31.10}$$

Any set of N^2 quantities C^i_j transforming in this fashion is a *mixed tensor*, i.e. it possesses both contravariant and covariant properties as is indicated by the two positions of its indices.

Similarly, the transformation law for a covariant tensor of rank 2 can be assembled from the law for covariant vectors.

The further generalization to tensors of higher rank should now be an obvious step. It will be sufficient to give one example. A^i_{jk} is a mixed tensor of rank 3, having both the covariant and contravariant properties indicated by the positions of its indices, if it transforms according to the equation

$$\overline{A}^{\,i}_{\ jk} = \frac{\partial \overline{x}^i}{\partial x^r} \frac{\partial x^s}{\partial \overline{x}^j} \frac{\partial x^t}{\partial \overline{x}^k} A^r_{st} \tag{31.11}$$

The components of a tensor can be given arbitrary values in any one frame and their values in any other frame are then uniquely determined by the transformation law. Consider the mixed second rank tensor whose components in the x^i-frame are δ^i_j, the Kronecker deltas ($\delta^i_j = 0, i \neq j$ and $\delta^i_j = 1, i = j$). The components in the $\overline{x}^i$-frame are $\overline{\delta}^{\,i}_{\ j}$, where

$$\begin{aligned}
\overline{\delta}^{\,i}_j &= \frac{\partial \overline{x}^i}{\partial x^k} \frac{\partial x^l}{\partial \overline{x}^j} \delta^k_l \\
&= \frac{\partial \overline{x}^i}{\partial x^k} \frac{\partial x^k}{\partial \overline{x}^j} \\
&= \frac{\partial \overline{x}^i}{\partial \overline{x}^j} \\
&= \delta^i_j \tag{31.12}
\end{aligned}$$

Thus this tensor has the same components in all frames and is called the *fundamental mixed tensor*. However, a second rank *covariant* tensor whose components in the x^i-frame are the Kronecker deltas (in this case denoted by δ_{ij}), has different components in other frames and is accordingly of no special interest.

It is reasonable to enquire at this stage why the distinction between covariant and contravariant tensors did not arise when the coordinate transformations were restricted to be orthogonal. Thus, suppose that A^i, B_i are contravariant and covariant vectors with respect to the orthogonal transformation (8.1). The inverse transformation has been shown to be equation (11.5) and it follows from these two equations that

$$\frac{\partial \overline{x}_i}{\partial x_j} = a_{ij}, \quad \frac{\partial x_i}{\partial \overline{x}_j} = a_{ji} \tag{31.13}$$

For the particular case of orthogonal transformations, therefore, equations (31.3) and (31.8) take the form

$$\overline{A}^{\,i} = a_{ij}A^j, \quad \overline{B}_i = a_{ij}B_j \tag{31.14}$$

It is clear that both types of vector transform in an identical manner and the distinction between them cannot, therefore, be maintained.

As in the case of the Cartesian tensors of Chapter 2, new tensors may be formed from known tensors by addition (or subtraction) and multiplication. Only tensors of the same rank and type may be added to yield new tensors. Thus, if A^i_{jk}, B^i_{jk} are components of tensors and we define the quantities C^i_{jk} by the equation

$$C^i_{jk} = A^i_{jk} + B^i_{jk} \tag{31.15}$$

then C^i_{jk} are the components of a tensor having the covariant and contravariant properties indicated by the position of its indices. However, A^i_j, B_{ij} cannot be added in this way to yield a tensor. Any two tensors may be multiplied to yield a new tensor. Thus, if A^i_j, B^k_{lm} are tensors and we define N^5 quantities C^{ik}_{jlm} by the equation

$$C^{ik}_{jlm} = A^i_j B^k_{lm} \tag{31.16}$$

these are the components of a fifth rank tensor having the covariant and contravariant properties indicated by its indices. The proofs of these statements are left for the reader to provide.

If a tensor is symmetric (or skew-symmetric) with respect to two of its superscripts or to two of its subscripts in any one frame, then it possesses this property in every frame. The method of proof is identical with that of the corresponding statement for Cartesian tensors given in section 10. However, if $A^i_j = A^j_i$ is true for all i, j when one reference frame is being employed, this equation will not, in general, be valid in any other frame. Thus, symmetry (or skew-symmetry) of a tensor with respect to a superscript and a subscript is not, in general, a covariant property. The tensor δ^i_j is exceptional in this respect.

Another result of great importance which may be established by the same argument we employed in the particular case of Cartesian tensors, is that an equation between tensors of the same type and rank is valid in all frames if it is valid in one. This implies that such tensor equations are covariant (i.e. are of invariable form) with respect to transformations between reference frames. The usefulness of tensors for our later work will be found to depend chiefly upon this property.

A symbol such as A^i_{jk} can be *contracted* by setting a superscript and a subscript to be the same letter. Thus A^i_{ji}, A^i_{ik} are the possible contractions of A^i_{jk} and each, by the repeated index summation convention, represents a sum. Since in the symbol A^i_{ji}, j alone is a free index, this entity has only N components. Similarly A^i_{ik} has N components. It will now be proved that, if A^i_{jk} is a tensor, its contractions are also tensors. Specifically, we shall prove that $B_j = A^i_{ji}$ is a covariant vector. For

$$\bar{B}_j = \bar{A}^i_{ji} = \frac{\partial \bar{x}^i}{\partial x^r} \frac{\partial x^s}{\partial \bar{x}^j} \frac{\partial x^t}{\partial \bar{x}^i} A^r_{st}$$

$$= \left(\frac{\partial x^t}{\partial \bar{x}^i} \frac{\partial \bar{x}^i}{\partial x^r} \right) \frac{\partial x^s}{\partial \bar{x}^j} A^r_{st}$$

$$= \frac{\partial x^t}{\partial x^r} \frac{\partial x^s}{\partial \bar{x}^j} A^r_{st}$$

$$= \delta^t_r \frac{\partial x^s}{\partial \bar{x}^j} A^r_{st}$$

$$= \frac{\partial x^s}{\partial \bar{x}^j} A^t_{st}$$

$$= \frac{\partial x^s}{\partial \bar{x}^j} B_s \tag{31.17}$$

establishing the result. This argument can obviously be generalized to yield the result that any contracted tensor is itself a tensor of rank two less than the tensor from which it has been derived and of the type indicated by the positions of its remaining free indices. In this connection it should be noted that, if A^i_{jk} is a tensor, A^i_{jj} is not, in general, a tensor; it is essential that the contraction be with respect to a superscript and a subscript and not with respect to two indices of the same kind.

If A^i_{jk}, B^r_s are tensors, the tensor $A^i_{jk} B^r_s$ is called the *outer product* of these two tensors. If this product is now contracted with respect to a superscript of one factor and a subscript of the other, e.g. $A^i_{jk} B^r_i$, the result is a tensor called an *inner product*.

32. The quotient theorem. Conjugate tensors

It has been remarked in the previous section that both the outer and inner products of two tensors are themselves tensors. Suppose, however, that it is known that a product of two factors is a tensor and that one of the factors is a tensor, can it be concluded that the other factor is also a tensor? We shall prove the following *quotient theorem*:

If the result of taking the product (outer or inner) of a given set of elements with a tensor of any specified type and arbitrary components is known to be a tensor, then the given elements are the components of a tensor.

It will be sufficient to prove the theorem true for a particular case, since the argument will easily be seen to be of general application. Thus, suppose the A^i_{jk} are N^3 quantities and it is to be established that these are the components of a tensor of the type indicated by the positions of the indices. Let B^r_s be a mixed tensor of rank 2 whose components can be chosen arbitrarily (in any one frame only of course) and suppose it is given that the inner product

$$A^i_{jk}B^k_s = C^i_{js} \tag{32.1}$$

is a tensor for all such B^r_s. All components have been assumed calculated with respect to the x-frame. Transforming to the $\bar{x}$-frame, the inner product is given to transform as a tensor and hence we have

$$A^{i*}_{jk}\overline{B}^k_s = \overline{C}^i_{js} \tag{32.2}$$

where A^{i*}_{jk} are the actual components replacing the A^i_{jk} when the reference frame is changed. Let $\overline{A}^i_{jk}$ be a set of elements *defined* in the $\bar{x}$-frame by equation (31.11). Since this is a tensor transformation equation, we know that the elements so defined will satisfy

$$\overline{A}^i_{jk}\overline{B}^k_s = \overline{C}^i_{js} \tag{32.3}$$

Subtracting equation (32.3) from (32.2), we obtain

$$(A^{i*}_{jk} - \overline{A}^i_{jk})\overline{B}^k_s = 0 \tag{32.4}$$

Since B^r_s has arbitrary components in the x-frame, its components in the $\bar{x}$-frame are also arbitrary and the components $\overline{B}_s^k$ can assume any convenient values. Thus, taking $\overline{B}_s^k = 1$ when $k = K$ and $\overline{B}_s^k = 0$ otherwise, equation (32.4) yields

$$A_{jK}^{i*} - \overline{A}_{jK}^i = 0$$

or
$$A_{jK}^{i*} = \overline{A}_{jK}^i \tag{32.5}$$

This being true for $K = 1, 2, \ldots, N$, we have quite generally

$$A_{jk}^{i*} = \overline{A}_{jk}^i \tag{32.6}$$

This implies that A_{jk}^i does transform as a tensor.

We will first give a very simple example of the application of this theorem. Let A^i be an arbitrary contravariant vector. Then

$$\delta_j^i A^j = A^i \tag{32.7}$$

and since the right-hand member of this equation is certainly a vector, by the quotient theorem δ_j^i is a tensor (as we have proved earlier).

As a second example, let g_{ij} be a symmetric covariant tensor and let $g = |g_{ij}|$ be the determinant whose elements are the tensor's components. We shall denote by G^{ij} the co-factor in this determinant of the element g_{ij}. Then, although G^{ij} is not a tensor, if $g \neq 0$, $G^{ij}/g = g^{ij}$ is a symmetric contravariant tensor. To prove this, we first observe that

$$g_{ij}G^{kj} = g\delta_i^k, \quad g_{ij}G^{ik} = g\delta_j^k \tag{32.8}$$

and hence, dividing by g,

$$g_{ij}g^{kj} = \delta_i^k, \quad g_{ij}g^{ik} = \delta_j^k \tag{32.9}$$

Now let A^i be an arbitrary contravariant vector and define the covariant vector B_i by the equation

$$B_i = g_{ik}A^k \tag{32.10}$$

Since $g \neq 0$, when the components of B_i are chosen arbitrarily, the corresponding components of A^i can always be calculated from this last equation, i.e. B_i is arbitrary with A^i. But

$$g^{ij}B_i = g^{ij}g_{ik}A^k = \delta_k^j A^k = A^j \tag{32.11}$$

having employed the second identity (32.9). It now follows by the quotient theorem that g^{ij} is a contravariant tensor. That it is symmetric follows from the circumstance that G^{ij} possesses this property. g_{ij}, g^{ij} are said to be *conjugate* to one another.

33. Covariant derivatives. Parallel displacement. Affine connection

In the earlier sections of this chapter, the algebra of tensors was established and it is now time to explain how the concepts of analysis can be introduced into the theory. Our space $\mathscr{S}_N$ has N dimensions, but is otherwise almost devoid of special characteristics. Nonetheless, it has so far been able to provide all the facilities required of a stage upon which the tensors are to play their roles. It will now be demonstrated, however, that additional features must be built into the structure of $\mathscr{S}_N$, before it can function as a suitable environment for the operations of tensor analysis.

It has been proved that, if ϕ is an invariant field, $\partial\phi/\partial x^i$ is a covariant vector. But, if a covariant vector is differentiated, the result is not a tensor. For, let A_i be such a vector, so that

$$\overline{A}_i = \frac{\partial x^k}{\partial \overline{x}^i} A_k \tag{33.1}$$

Differentiating both sides of this equation with respect to $\overline{x}^j$, we obtain

$$\frac{\partial \overline{A}_i}{\partial \overline{x}^j} = \frac{\partial x^k}{\partial \overline{x}^i} \frac{\partial x^l}{\partial \overline{x}^j} \frac{\partial A_k}{\partial x^l} + \frac{\partial^2 x^k}{\partial \overline{x}^i \partial \overline{x}^j} A_k \tag{33.2}$$

The presence of the second term of the right-hand member of this equation reveals that $\partial A_i/\partial x^j$ does not transform as a tensor. However, this fact can be arrived at in a more revealing manner as follows:

Let P, P′ be the neighbouring points $x^i, x^i + dx^i$ and let $A_i, A_i + dA_i$ be the vectors of a covariant vector field associated with these points respectively. The transformation laws for these two vectors will be different, since the coefficients of a tensor transformation law vary from point to point in $\mathscr{S}_N$. It follows that the difference of these two vectors, namely dA_i, is not a vector. However,

$$dA_i = \frac{\partial A_i}{\partial x^j} \, dx^j \tag{33.3}$$

and, since dx^j is a vector, if $A_{i,\,j}$ were a tensor, dA_i would be a vector. $A_{i,\,j}$ cannot be a tensor, therefore. The source of the difficulty is now apparent. To define $A_{i,\,j}$ it is necessary to compare the values assumed by the vector field A_i at two neighbouring, but distinct, points and such a comparison cannot lead to a tensor. If, however, this procedure could be replaced by another, involving the comparison of two vectors defined at the same point, the modified equation (33.3) would be expected to be a tensor equation featuring a new form of derivative which is a tensor. This leads us quite naturally to the concept of *parallel displacement*.

Suppose that the vector A_i is displaced from the point P, at which it is defined, to the neighbouring point P′, *without change in magnitude or direction*, so that it may be thought of as being the same vector now defined at the neighbouring point. The phrase in italics has no precise meaning in $\mathscr{S}_N$ as yet, for we have not defined the magnitude or the direction of a vector in this space. However, in the particular case when $\mathscr{S}_N$ is Euclidean and rectangular axes are being employed, this phrase is, of course, interpreted as requiring that the displaced vector shall possess the same components as the original vector. But even in $\mathscr{E}_N$, if curvilinear coordinates are being used, the directions of the curvilinear axes at the point P′

will, in general, be different from their directions at P and, as a consequence, the components of the displaced vector will not be identical with its components before the displacement. In $\mathscr{S}_N$, therefore, components of the displaced vector will be denoted by $A_i + \delta A_i$. This vector can now be compared with the field vector $A_i + dA_i$ at the same point P'. Since the two vectors are defined at the same point, their difference is a vector at this point, i.e. $dA_i - \delta A_i$, is a vector. The modified equation (33.3) is accordingly expected to be of the form

$$dA_i - \delta A_i = A_{i;j}\, dx^j \qquad (33.4)$$

where $A_{i;j}$ is the appropriate replacement for $A_{i,j}$. Since dx^j is an arbitrary vector and the left-hand member of equation (33.4) is known to be a vector, $A_{i;j}$ will, by the quotient theorem, be a covariant tensor. It will be termed the *covariant derivative* of A_i. Thus, the problem of defining a tensor derivative has been re-expressed as the problem of defining parallel displacement (infinitesimal) of a vector.

We are at liberty to define the parallel displacement of A_i from P to P' in any way we shall find convenient. However, to avoid confusion, it is necessary that the definition we accept shall be in conformity with that adopted in $\mathscr{E}_N$, which is a special case of $\mathscr{S}_N$. Suppose, therefore, that our $\mathscr{S}_N$ is Euclidean and that y^i are rectangular Cartesian coordinates in this space. Let B_i be the components of the vector field A_i with respect to these rectangular axes. Then

$$A_i = \frac{\partial y^j}{\partial x^i} B_j, \quad B_i = \frac{\partial x^j}{\partial y^i} A_j \qquad (33.5)$$

If the parallel displacement of the vector A_i to the point P' is now carried out, its Cartesian components B_i will not change, i.e. $\delta B_i = 0$. Hence, from the first of equations (33.5), we obtain

$$\delta A_i = \delta\left(\frac{\partial y^j}{\partial x^i} B_j\right) = \delta\left(\frac{\partial y^j}{\partial x^i}\right) B_j$$

$$= \frac{\partial^2 y^j}{\partial x^i \partial x^k}\, dx^k B_j \qquad (33.6)$$

Substituting for B_j into this equation from the second of equations (33.5), we find that

$$\delta A_i = \Gamma_{ik}^l A_l dx^k \qquad (33.7)$$

where

$$\Gamma_{ik}^l = \frac{\partial^2 y^j}{\partial x^i \partial x^k} \frac{\partial x^l}{\partial y^j} \qquad (33.8)$$

This shows that, in $\mathscr{E}_N$, the δA_i are bilinear forms in the A_l and dx^k. In $\mathscr{S}_N$, we shall accordingly *define* the δA_i by the equation (33.7), determining the N^3 quantities Γ_{ik}^l arbitrarily at every point of $\mathscr{S}_N$.* This set of quantities Γ_{ik}^l is called an *affinity*

---

* Subject to the requirement that the Γ_{ik}^l are continuous functions of the x^i and possess continuous partial derivatives to the order necessary to validate all later arguments.

and specifies an *affine connection* between the points of $\mathscr{S}_N$. A space which is affinely connected possesses sufficient structure to permit the operations of tensor analysis to be carried out within it.

For we can now write

$$\mathrm{d}A_i - \delta A_i = \frac{\partial A_i}{\partial x^j}\mathrm{d}x^j - \Gamma_{ij}^k A_k \mathrm{d}x^j$$

$$= \left(\frac{\partial A_i}{\partial x^j} - \Gamma_{ij}^k A_k\right)\mathrm{d}x^j \tag{33.9}$$

But, as we have already explained, the left-hand member of this equation is a vector for arbitrary $\mathrm{d}x^j$ and hence it follows that

$$A_{i;j} = \frac{\partial A_i}{\partial x^j} - \Gamma_{ij}^k A_k \tag{33.10}$$

is a covariant tensor, the covariant derivative of A_i.

It will be observed from equation (33.10) that, if the components of the affinity all vanish over some region of $\mathscr{S}_N$, the covariant and partial derivatives are identical over this region. However, this will only be the case in the particular reference frame being employed. In any other frame the components of the affinity will, in general, be non-zero and the distinction between the two derivatives will be maintained. In tensor equations which are to be valid in every frame, therefore, only covariant derivatives may appear, even if it is possible to find a frame relative to which the affinity vanishes.

We have stated earlier that, when defining an affine connection, the components of an affinity may be chosen arbitrarily. To be precise, a coordinate frame must first be selected in $\mathscr{S}_N$ and the choice of the components of the affinity is then arbitrary within this frame. However, when these have been determined, the components of the affinity with respect to any other frame are, as for tensors, completely fixed by a transformation law. We now proceed to obtain this transformation law for affinities.

34. Transformation of an affinity

The manner in which each of the quantities occurring in equation (33.10) transforms is known, with the exception of the affinity Γ_{ij}^k. The transformation law for this affinity can accordingly be deduced by transformation of this equation. Relative to the $\bar{x}$-frame, the equation is written

$$\bar{A}_{i;j} = \frac{\partial \bar{A}_i}{\partial \bar{x}^j} - \bar{\Gamma}_{ij}^k \bar{A}_k \tag{34.1}$$

Since A_i, $A_{i;j}$ are tensors,

$$\bar{A}_i = \frac{\partial x^r}{\partial \bar{x}^i} A_r \tag{34.2}$$

$$\overline{A}_{i;j} = \frac{\partial x^s}{\partial \overline{x}^i} \frac{\partial x^t}{\partial \overline{x}^j} A_{s;t} \tag{34.3}$$

Substituting in equation (34.1), we obtain

$$\frac{\partial x^s}{\partial \overline{x}^i} \frac{\partial x^t}{\partial \overline{x}^j} A_{s;t} = \frac{\partial x^r}{\partial \overline{x}^i} \frac{\partial x^u}{\partial \overline{x}^j} \frac{\partial A_r}{\partial x^u} + \frac{\partial^2 x^r}{\partial \overline{x}^i \partial \overline{x}^j} A_r - \overline{\Gamma}^k_{ij} \frac{\partial x^r}{\partial \overline{x}^k} A_r \tag{34.4}$$

Employing equation (33.10) to substitute for $A_{s;t}$ and cancelling a pair of identical terms from the two members of equation (34.4), this equation reduces to

$$-\frac{\partial x^s}{\partial \overline{x}^i} \frac{\partial x^t}{\partial \overline{x}^j} \Gamma^r_{st} A_r = \frac{\partial^2 x^r}{\partial \overline{x}^i \partial \overline{x}^j} A_r - \overline{\Gamma}^k_{ij} \frac{\partial x^r}{\partial \overline{x}^k} A_r \tag{34.5}$$

Since A_r is an arbitrary vector, we can equate coefficients of A_r from the two members of this equation to obtain

$$\overline{\Gamma}^k_{ij} \frac{\partial x^r}{\partial \overline{x}^k} = \frac{\partial x^s}{\partial \overline{x}^i} \frac{\partial x^t}{\partial \overline{x}^j} \Gamma^r_{st} + \frac{\partial^2 x^r}{\partial \overline{x}^i \partial \overline{x}^j} \tag{34.6}$$

Multiplying both sides of this equation by $\partial \overline{x}^l/\partial x^r$ and using the result

$$\frac{\partial \overline{x}^l}{\partial x^r} \frac{\partial x^r}{\partial \overline{x}^k} = \frac{\partial \overline{x}^l}{\partial \overline{x}^k} = \delta^l_k \tag{34.7}$$

yields finally

$$\overline{\Gamma}^l_{ij} = \frac{\partial \overline{x}^l}{\partial x^r} \frac{\partial x^s}{\partial \overline{x}^i} \frac{\partial x^t}{\partial \overline{x}^j} \Gamma^r_{st} + \frac{\partial \overline{x}^l}{\partial x^r} \frac{\partial^2 x^r}{\partial \overline{x}^i \partial \overline{x}^j} \tag{34.8}$$

which is the transformation law for an affinity.

It should be noted that, were it not for the presence of the second term in the right-hand member of equation (34.8), Γ^k_{ij} would transform as a tensor of the third rank having the covariant and contravariant characteristics suggested by the positions of its indices. Thus, the transformation law is linear in the components of an affinity but is not homogeneous like a tensor transformation law. This has the consequence that, if all the components of an affinity are zero relative to one frame, they are not necessarily zero relative to another frame. However, in general, there will be no frame in which the components of an affinity vanish over a region of $\mathcal{S}_N$, though it will be proved that, provided the affinity is symmetric, it is always possible to find a frame in which the components all vanish at some particular point (see section 39).

Suppose Γ^k_{ij}, $\Gamma^k_{ij}{}^*$ are two affinities defined over a region of $\mathcal{S}_N$. Writing down their transformation laws and subtracting one from the other, it is immediate that

$$\overline{\Gamma}^k_{ij} - \overline{\Gamma}^k_{ij}{}^* = \frac{\partial \overline{x}^k}{\partial x^r} \frac{\partial x^s}{\partial \overline{x}^i} \frac{\partial x^t}{\partial \overline{x}^j} (\Gamma^r_{st} - \Gamma^r_{st}{}^*) \tag{34.9}$$

i.e. the difference of two affinities is a tensor. However, the sum of two affinities is neither a tensor nor an affinity. It is left as an exercise for the reader to show, similarly, that the sum of an affinity Γ^k_{ij} and a tensor A^k_{ij} is an affinity.

If Γ^k_{ij} is symmetric with respect to its subscripts in one frame, it is symmetric in every frame. For, from equation (34.8),

$$
\begin{aligned}
\bar{\Gamma}^k_{ji} &= \frac{\partial \bar{x}^k}{\partial x^r} \frac{\partial x^s}{\partial \bar{x}^j} \frac{\partial x^t}{\partial \bar{x}^i} \Gamma^r_{st} + \frac{\partial \bar{x}^k}{\partial x^r} \frac{\partial^2 x^r}{\partial \bar{x}^j \partial \bar{x}^i} \\
&= \frac{\partial \bar{x}^k}{\partial x^r} \frac{\partial x^t}{\partial \bar{x}^i} \frac{\partial x^s}{\partial \bar{x}^j} \Gamma^r_{ts} + \frac{\partial \bar{x}^k}{\partial x^r} \frac{\partial^2 x^r}{\partial \bar{x}^i \partial \bar{x}^j} \\
&= \bar{\Gamma}^k_{ij}
\end{aligned}
\tag{34.10}
$$

where, at the first step, we have put $\Gamma^r_{ts} = \Gamma^r_{st}$.

35. Covariant derivatives of tensors

In this section, we shall extend the process of covariant differentiation to tensors of all ranks and types.

Consider first an invariant field V. When V suffers parallel displacement from P to P′, its value will be taken to be unaltered, i.e. $\delta V = 0$ in all frames. Hence

$$
dV - \delta V = \frac{\partial V}{\partial x^i} dx^i
\tag{35.1}
$$

is the counterpart for an invariant of equation (33.4). It follows that

$$
V_{;i} = V_{,i}
\tag{35.2}
$$

i.e. the covariant derivative of an invariant is identical with its partial derivative or gradient.

Now let B^i be a contravariant vector field and A_i an arbitrary covariant vector. Then $A_i B^i$ is an invariant and, when parallel displaced from P to P′, remains unchanged in value. Thus

$$
\delta(A_i B^i) = 0
$$

or

$$
\delta A_i B^i + A_i \delta B^i = 0
$$

and hence, by equation (33.7),

$$
A_k \delta B^k = -\Gamma^k_{ij} A_k dx^j B^i
\tag{35.3}
$$

But, since the A_k are arbitrary, their coefficients in the two members of this equation can be equated to yield

$$
\delta B^k = -\Gamma^k_{ij} B^i dx^j
\tag{35.4}
$$

This equation defines the parallel displacement of a contravariant vector. The covariant derivative of the vector is now deduced as before: thus

$$
dB^k - \delta B^k = \left(\frac{\partial B^k}{\partial x^j} + \Gamma^k_{ij} B^i \right) dx^j
\tag{35.5}
$$

and since dx^j is an arbitrary vector and $dB^k - \delta B^k$ is then known to be a vector,

$$B^k_{;j} = \frac{\partial B^k}{\partial x^j} + \Gamma^k_{ij} B^i \tag{35.6}$$

is a tensor called the covariant derivative of B^k.

Similarly, if A^i_j is a tensor field, we consider the parallel displacement of the invariant $A^i_j B_i C^j$, where B_i, C^j are arbitrary vectors. Then, from

$$\delta(A^i_j B_i C^j) = 0 \tag{35.7}$$

and equations (33.7) and (35.4), we deduce that

$$\delta A^i_j = \Gamma^l_{jk} A^i_l dx^k - \Gamma^i_{lk} A^l_j dx^k \tag{35.8}$$

It now follows that

$$A^i_{j;k} = \frac{\partial A^i_j}{\partial x^k} - \Gamma^l_{jk} A^i_l + \Gamma^i_{lk} A^l_j \tag{35.9}$$

is the covariant derivative required.

The rule for finding the covariant derivative of any tensor will now be plain from examination of equation (35.9), viz., the appropriate partial derivative is first written down and this is then followed by 'affinity terms'; the 'affinity terms' are obtained by writing down an inner product of the affinity and the tensor with respect to each of its indices in turn, prefixing a positive sign when the index is contravariant and a negative sign when it is covariant.

Applying this rule to the tensor field whose components at every point are those of the fundamental tensor δ^i_j, it will be found that

$$\delta^i_{j;k} = \Gamma^i_{rk} \delta^r_j - \Gamma^r_{jk} \delta^i_r = \Gamma^i_{jk} - \Gamma^i_{jk} = 0 \tag{35.10}$$

Thus, the fundamental tensor behaves like a constant in covariant differentiation.

Finally, in this section, we shall demonstrate that the ordinary rules for the differentiation of sums and products apply to the process of covariant differentiation.

The right-hand member of equation (35.9) being linear in the tensor A^i_j, it follows immediately that if

$$C^i_j = A^i_j + B^i_j \tag{35.11}$$

then

$$C^i_{j;k} = A^i_{j;k} + B^i_{j;k} \tag{35.12}$$

Now suppose that

$$C^i = A^i_j B^j \tag{35.13}$$

Then

$$C^i_{;k} = \frac{\partial C^i}{\partial x^k} + \Gamma^i_{rk} C^r$$

$$= \frac{\partial}{\partial x^k}(A^i_j B^j) + \Gamma^i_{rk} A^r_j B^j$$

$$= \left(\frac{\partial A_j^i}{\partial x^k} + \Gamma_{rk}^i A_j^r - \Gamma_{jk}^r A_r^i\right)B^j + \left(\frac{\partial B^j}{\partial x^k} + \Gamma_{rk}^j B^r\right)A_j^i$$

$$= A_{j;k}^i B^j + B_{;k}^j A_j^i \tag{35.14}$$

which is the ordinary rule for the differentiation of a product.

36. The Riemann–Christoffel curvature tensor

If a rectangular Cartesian coordinate frame is chosen in a Euclidean space $\mathscr{E}_N$ and if A^i are the components of a vector defined at a point Q with respect to this frame, then $\delta A^i = 0$ for an arbitrary small parallel displacement of the vector from Q. This being true for arbitrary A^i, it follows from equation (35.4) that $\Gamma_{jk}^i = 0$ with respect to this frame at every point of $\mathscr{E}_N$. Suppose C is a closed curve passing through Q and that A^i makes one complete circuit of C, being parallel displaced over each element of the path. Then its components remain unchanged throughout the motion and hence, if $A^i + \Delta A^i$ denotes the vector upon its return to Q,

$$\Delta A^i = 0 \tag{36.1}$$

Since ΔA^i is the difference between two vectors both defined at Q, it is itself a vector and equation (36.1) will therefore be a vector equation true in all frames. Thus, in $\mathscr{E}_N$, parallel displacement of a vector through one circuit of a closed curve leaves the vector unchanged.

If, however, A^i is defined at a point Q in an affinely connected space $\mathscr{S}_N$, not necessarily Euclidean, it will no longer be possible, in general, to choose a coordinate frame for which the components of the affinity vanish at every point. As a consequence, if A^i is parallel displaced around C, its components will vary and it is no longer permissible to suppose that upon its return to Q it will be unchanged, i.e. $\Delta A^i \neq 0$. We shall now calculate ΔA^i when A^i is parallel displaced around a small circuit C enclosing the point P having coordinates x^i (Fig. 7) at which it is initially defined.

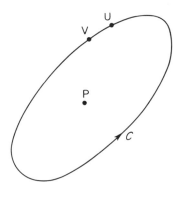

FIG. 7

Let U be any point on this curve and let $x^i + \xi^i$ be its coordinates, the ξ^i being small quantities. V is a point on C near to U and having coordinates $x^i + \xi^i + d\xi^i$. When A^i is displaced from U to V, its components undergo a change

$$\delta A^i = -\Gamma^i_{jk} A^j d\xi^k \qquad (36.2)$$

where Γ^i_{jk} and A^j are to be computed at U. Considering the small displacement from P to U and employing Taylor's theorem, the value of Γ^i_{jk} at U is seen to be

$$\Gamma^i_{jk} + \frac{\partial \Gamma^i_{jk}}{\partial x^l} \xi^l \qquad (36.3)$$

to the first order in the ξ^l. In this expression, the affinity and its derivative are to be computed at P. A^j in equation (36.2) represents the vector after its parallel displacement from P to U, i.e. it is

$$A^j - \Gamma^j_{rl} A^r \xi^l \qquad (36.4)$$

where A^j, A^r and Γ^j_{rl} are all to be calculated at the point P. To the first order in ξ^l therefore, equation (36.2) may be written

$$\delta A^i = -\left[\Gamma^i_{jk} A^j + \left(A^j \frac{\partial \Gamma^i_{jk}}{\partial x^l} - \Gamma^i_{jk} \Gamma^j_{rl} A^r \right) \xi^l \right] d\xi^k \qquad (36.5)$$

Integrating around C, it will be found that

$$\Delta A^i = -\Gamma^i_{jk} A^j \oint_C d\xi^k + \left(\Gamma^i_{rk} \Gamma^r_{jl} - \frac{\partial \Gamma^i_{jk}}{\partial x^l} \right) A^j \oint_C \xi^l d\xi^k \qquad (36.6)$$

where the dummy indices j, r have been interchanged in the final term of the right-hand member of equation (36.5).

Now

$$\oint_C d\xi^k = \Delta \xi^k = 0 \qquad (36.7)$$

Also

$$\oint_C d(\xi^l \xi^k) = \Delta(\xi^l \xi^k) = 0 \qquad (36.8)$$

so that

$$\oint_C \xi^l d\xi^k = -\oint_C \xi^k d\xi^l \qquad (36.9)$$

implying that the left-hand member of this equation is skew-symmetric in l and k. Since ξ^l, $d\xi^k$ are vectors, it is also a tensor. Denoting it by α^{kl}, we have

$$\alpha^{kl} = \tfrac{1}{2} \oint_C (\xi^l d\xi^k - \xi^k d\xi^l) \qquad (36.10)$$

and equation (36.6) then reduces to the form

$$\Delta A^i = \left(\Gamma^i_{rk} \Gamma^r_{jl} - \frac{\partial \Gamma^i_{jk}}{\partial x^l} \right) A^j \alpha^{kl} \tag{36.11}$$

Apart from its property of skew-symmetry, α^{kl} is arbitrary. Nonetheless, since it is not completely arbitrary, the quotient theorem (section 32) cannot be applied directly to deduce that the contents of the bracket in equation (36.11) constitute a tensor. In fact, this expression is not a tensor. However, it is easy to prove that, if X^i_{kl} is skew-symmetric with respect to k, l and if Y^i, defined by the equation

$$Y^i = X^i_{kl} \alpha^{kl} \tag{36.12}$$

is a vector for arbitrary skew-symmetric tensors α^{kl}, then X^i_{kl} is also a tensor.

To prove this, let β^{kl} be an arbitrary symmetric tensor. Then the components of the tensor

$$\gamma^{kl} = \alpha^{kl} + \beta^{kl} \tag{36.13}$$

are completely arbitrary, for, assuming $k < l$,

$$\gamma^{kl} = \alpha^{kl} + \beta^{kl}, \quad \gamma^{lk} = -\alpha^{kl} + \beta^{kl} \tag{36.14}$$

and it follows that the values of γ^{kl}, γ^{lk} can be chosen arbitrarily and then

$$\alpha^{kl} = \tfrac{1}{2}(\gamma^{kl} - \gamma^{lk}), \quad \beta^{kl} = \tfrac{1}{2}(\gamma^{kl} + \gamma^{lk}) \tag{36.15}$$

i.e., it is only necessary to fix the values of α^{kl}, β^{kl} in the cases $k < l$ in order that the γ^{kl} shall assume any specified values over the complete range of its superscripts, with the exception of the cases when two superscripts are equal. If the superscripts are equal, $\alpha^{kl} = 0$ and $\gamma^{kl} = \beta^{kl}$. But these β^{kl} are also arbitrary and hence so again are the γ^{kl} with equal superscripts.

Since β^{kl} is symmetric and X^i_{kl} is skew-symmetric,

$$X^i_{kl} \beta^{kl} = 0 \tag{36.16}$$

Adding equations (36.12) and (36.16), we obtain therefore

$$X^i_{kl} \gamma^{kl} = Y^i \tag{36.17}$$

But γ^{kl} is an arbitrary tensor and hence, by the quotient theorem, X^i_{kl} is a tensor.

The multiplier of α^{kl} in equation (36.11) is not skew-symmetric in k, l. However, it can be made so as follows: Interchange the dummy indices k, l in this equation to obtain

$$\Delta A^i = \left(\Gamma^i_{rl} \Gamma^r_{jk} - \frac{\partial \Gamma^i_{jl}}{\partial x^k} \right) A^j \alpha^{lk} \tag{36.18}$$

Adding equations (36.11) and (36.18) and noting that $\alpha^{kl} = -\alpha^{lk}$, it will be found that

$$\Delta A^i = \tfrac{1}{2} \left(\Gamma^i_{rk} \Gamma^r_{jl} - \Gamma^i_{rl} \Gamma^r_{jk} + \frac{\partial \Gamma^i_{jl}}{\partial x^k} - \frac{\partial \Gamma^i_{jk}}{\partial x^l} \right) A^j \alpha^{kl} \tag{36.19}$$

The bracketed expression is now skew-symmetric in k, l and hence

$$\left(\Gamma^i_{rk}\Gamma^r_{jl} - \Gamma^i_{rl}\Gamma^r_{jk} + \frac{\partial\Gamma^i_{jl}}{\partial x^k} - \frac{\partial\Gamma^i_{jk}}{\partial x^l}\right)A^j \tag{36.20}$$

is a tensor. A^j being arbitrary, it follows that

$$B^i_{jkl} = \Gamma^i_{rk}\Gamma^r_{jl} - \Gamma^i_{rl}\Gamma^r_{jk} + \frac{\partial\Gamma^i_{jl}}{\partial x^k} - \frac{\partial\Gamma^i_{jk}}{\partial x^l} \tag{36.21}$$

is a tensor. It is the *Riemann–Christoffel curvature tensor*.

Equation (36.19) can now be written

$$\Delta A^i = \tfrac{1}{2}B^i_{jkl}A^j\alpha^{kl} \tag{36.22}$$

If B^i_{jkl} is contracted with respect to the indices i and l, the resulting tensor is called the *Ricci tensor* and is denoted by R_{jk}. Thus

$$R_{jk} = B^i_{jki} \tag{36.23}$$

This tensor has an important role to play in Einstein's theory of gravitation. Since B^i_{jkl} is skew-symmetric with respect to the indices k and l, its contraction with respect to i and k yields only the Ricci tensor again in the form $-R_{jl}$. However, contraction with respect to the indices i and j yields another second rank tensor, viz.

$$S_{kl} = B^i_{ikl} = \frac{\partial\Gamma^i_{il}}{\partial x^k} - \frac{\partial\Gamma^i_{ik}}{\partial x^l} \tag{36.24}$$

37. Metrical connection. Raising and lowering indices

In this section we shall further particularize our space $\mathscr{S}_N$ by supposing it to be Riemannian. That is, we shall suppose that a 'distance' or *interval* ds between two neighbouring points $x^i, x^i + dx^i$ is defined by the equation

$$ds^2 = g_{ij}dx^i dx^j \tag{37.1}$$

where the N^2 coefficients g_{ij} are specified in some coordinate frame at every point of $\mathscr{S}_N$. It will be assumed, without loss of generality, that the g_{ij} are symmetric. Such a relationship between all pairs of adjacent points is called a *metrical connection* and the expression (37.1) for ds^2 is termed the *metric*.

For any two neighbouring points, ds will be regarded as an invariant associated with them and the g_{ij} must accordingly transform so that this shall be so. Since $dx^i dx^j$ is an arbitrary symmetric tensor, g_{ij} is symmetric and ds^2 is an invariant, it follows by a modified quotient theorem similar to the one proved in section 36 that g_{ij} is a tensor. It is called the *fundamental covariant tensor*. The contravariant tensor which is conjugate to g_{ij} (see section 32), viz. g^{ij}, is termed the *fundamental contravariant tensor*. This exists only if $g = |g_{ij}| \neq 0$, which we accordingly assume to be the case.

In the case when $\mathscr{R}_N$ is Euclidean, rectangular Cartesian coordinates y^i can be defined and, in such a frame, $g_{ij} = \delta_{ij}$. Consider a contravariant vector having components A^i in a general curvilinear x-frame and components B^i in the y-frame. In the Cartesian frame, covariant and contravariant vectors are indistinguishable, so that it is natural to define covariant components for the vector by the equation

$$B_i = B^i \qquad (37.2)$$

In the x-frame, let A_i be the components of the covariant vector B_i. Then

$$A_i = g_{ij}A^j \qquad (37.3)$$

This follows since (i) it is a tensor equation and (ii) it is valid in the y-frame in which it takes the form

$$B_i = \delta_{ij}B^j = B^i \qquad (37.4)$$

If $\mathscr{R}_N$ is not Euclidean, equation (37.3) is taken to define the covariant components of a vector whose contravariant components are A^i. This process of converting the contravariant components of a vector into covariant components is termed *lowering the index*.

If B_i is a covariant vector, its contravariant expression is determined by *raising the index* with the aid of the fundamental contravariant vector. Thus

$$B^i = g^{ij}B_j \qquad (37.5)$$

For the notation to be consistent, it is necessary that if an index is first lowered and then raised, the original vector should again be obtained. This is seen to be the case for, if A_i is formed from A^i (equation (37.3)), the result of raising its subscript is (equation (37.5))

$$g^{ij}A_j = g^{ij}g_{jk}A^k = \delta^i_k A^k = A^i \qquad (37.6)$$

where equations (32.9) have been used in the reduction. Similarly, if an index is first raised and then lowered, the original covariant vector is reproduced.

Any index of a tensor can now be raised or lowered in the obvious way. Thus, if A^{ij}_k is a tensor, we define

$$A^{i\cdot}_{jk} = g_{jr}A^{ir}_k \qquad (37.7)$$

To allow for the possibility that indices may be raised or lowered during a calculation, it will be convenient to displace the subscripts to the right of the superscripts. It is also often helpful to keep a record of these operations by placing a dot in the gap resulting from the raising or lowering of an index. These conventions are illustrated in equation (37.7).

Suppose an index of the fundamental tensor g_{ij} is raised. The result is

$$g^k_{\cdot j} = g^{ki}g_{ij} = \delta^k_j \qquad (37.8)$$

i.e. the mixed fundamental tensor. The same tensor results when an index of g^{ij} is lowered. If both subscripts of g_{ij} are raised, the result is

$$g^{ri}g^{sj}g_{ij} = g^{ri}\delta_i^s = g^{rs} \tag{37.9}$$

Our notation is entirely consistent, therefore, and g_{ij}, g^{ij}, δ_j^i are taken to be the covariant, contravariant and mixed components respectively of a single fundamental tensor.

Consider the inner product of two vectors A^i, B_i. We have

$$\begin{aligned} A^i B_i &= g^{ij} A_j B_i \\ &= A_j g^{ij} B_i \\ &= A_j B^j \\ &= A_i B^i \end{aligned} \tag{37.10}$$

It is clear that the dummy index occurring in the expression for an inner product can be raised in one factor and lowered in the other without affecting the result. This is obviously valid for the inner product of any pair of tensors.

38. Scalar products. Magnitudes of vectors

In $\mathcal{R}_N$ the magnitude of the displacement vector dx^i is taken to be ds as given by equation (37.1). If A^i is any other contravariant vector, it may be represented as a displacement vector and then its *magnitude* is the invariant A, where

$$A^2 = g_{ij} A^i A^j \tag{38.1}$$

This equation is accordingly taken to define the magnitude of A^i.

Raising and lowering the dummy indices in equation (38.1), we obtain the equivalent result

$$A^2 = g^{ij} A_i A_j \tag{38.2}$$

It is natural to assume that the associated vectors A_i, A^i have equal magnitudes and hence A is also taken to be the magnitude of A_i. Equation (38.2) indicates how this can be calculated directly from A_i.

Since $g_{ij}A^j = A_i$ and $g^{ij}A_j = A^i$, equations (38.1), (38.2) are also both seen to be equivalent to the equation

$$A^2 = A_i A^i \tag{38.3}$$

The *scalar product* of two vectors $\mathbf{A}$, $\mathbf{B}$ is defined to be the invariant

$$\mathbf{A} \cdot \mathbf{B} = A_i B^i = A^i B_i = g_{ij} A^i B^j = g^{ij} A_i B_j \tag{38.4}$$

It will be noted that

$$A^2 = \mathbf{A} \cdot \mathbf{A} \tag{38.5}$$

By analogy with $\mathscr{E}_3$, we now define the angle θ between two vectors $\mathbf{A}$, $\mathbf{B}$ to be such that

$$AB\cos\theta = \mathbf{A} \cdot \mathbf{B} \tag{38.6}$$

i.e.
$$\cos\theta = \frac{A^i B_i}{\sqrt{[(A^j A_j)(B^k B_k)]}} \tag{38.7}$$

If $\theta = \frac{1}{2}\pi$, the vectors are said to be *orthogonal* and

$$A^i B_i = 0 \tag{38.8}$$

39. Geodesic frame. Christoffel symbols

It is always possible to choose a coordinate frame in which the components of the metric tensor are stationary at an assigned point, i.e. $\partial g_{ij}/\partial x^k$ all vanish at the point. Such a frame is said to be *geodesic* at the point.

For, if $x^i = a^i$ are the coordinates of a point A in an x-frame, suppose we transform to a new frame by the quadratic transformation

$$x^i = a^i + \bar{x}^i + \tfrac{1}{2}a^i_{jk}\bar{x}^j\bar{x}^k \tag{39.1}$$

where $\bar{x}^i = 0$ at A and the constant coefficients a^i_{jk} are symmetric in the indices j, k (no loss of generality). Differentiation yields the equations

$$\frac{\partial x^i}{\partial \bar{x}^j} = \delta^i_j + a^i_{jk}\bar{x}^k, \quad \frac{\partial^2 x^i}{\partial \bar{x}^j \partial \bar{x}^k} = a^i_{jk} \tag{39.2}$$

In particular, at the point A

$$\frac{\partial x^i}{\partial \bar{x}^j} = \delta^i_j, \quad \frac{\partial^2 x^i}{\partial \bar{x}^j \partial \bar{x}^k} = a^i_{jk} \tag{39.3}$$

The transformation equation for the metric tensor is

$$\bar{g}_{ij} = \frac{\partial x^r}{\partial \bar{x}^i}\frac{\partial x^s}{\partial \bar{x}^j}g_{rs} \tag{39.4}$$

Differentiating with respect to $\bar{x}^k$, we find that

$$\frac{\partial \bar{g}_{ij}}{\partial \bar{x}^k} = \frac{\partial^2 x^r}{\partial \bar{x}^k \partial \bar{x}^i}\frac{\partial x^s}{\partial \bar{x}^j}g_{rs} + \frac{\partial x^r}{\partial \bar{x}^i}\frac{\partial^2 x^s}{\partial \bar{x}^k \partial \bar{x}^j}g_{rs} + \frac{\partial x^r}{\partial \bar{x}^i}\frac{\partial x^s}{\partial \bar{x}^j}\frac{\partial g_{rs}}{\partial x^t}\frac{\partial x^t}{\partial \bar{x}^k} \tag{39.5}$$

Substituting from equations (39.3), it follows that at the point A,

$$\frac{\partial \bar{g}_{ij}}{\partial \bar{x}^k} = a^r_{ik}g_{rj} + a^s_{jk}g_{is} + \frac{\partial g_{ij}}{\partial x^k} \tag{39.6}$$

Suppose, if possible, the coefficients a^i_{jk} are chosen so that $\partial \bar{g}_{ij}/\partial \bar{x}^k = 0$. Then, writing $a^r_{ik}g_{rj} = a_{ikj}$, it is necessary that

$$a_{ikj} + a_{jki} = -\partial g_{ij}/\partial x^k \tag{39.7}$$

Cyclically permuting the indices i, j, k, two further equations are obtained, viz.

$$a_{jik} + a_{kij} = -\partial g_{jk}/\partial x^i \tag{39.8}$$

$$a_{kji} + a_{ijk} = -\partial g_{ki}/\partial x^j \tag{39.9}$$

Since a_{ijk} is symmetric in the indices i, j, by adding the last pair of equations and subtracting equation (39.7), we get

$$a_{ijk} = -\tfrac{1}{2}\left(\frac{\partial g_{jk}}{\partial x^i} + \frac{\partial g_{ki}}{\partial x^j} - \frac{\partial g_{ij}}{\partial x^k}\right) = -[ij, k] \tag{39.10}$$

$[ij, k]$ is called the *Christoffel symbol of the first kind*. It is now easily verified that the condition (39.7) is satisfied provided a_{ijk} is defined by equation (39.10) and, hence, that $\partial \bar{g}_{ij}/\partial \bar{x}^k = 0$ at A.

$[ij, k]$ is not a tensor, but its indices invariably behave like subscripts in any formula in which it occurs. It is symmetric in the indices i, j.

We now deduce that

$$a_{ij}^k = g^{kr} a_{ijr} = -g^{kr}[ij, r] = -\{_i{}^k{}_j\} \tag{39.11}$$

where $\{_i{}^k{}_j\}$ is *Christoffel's symbol of the second kind*. It is clearly obtained from the symbol of the first kind $[ij, k]$ by raising the final index k. It, also, is not a tensor, but i, j always behave as subscripts and k behaves as a superscript; it is symmetric in i and j.

Suppose we next transform from the $\bar{x}$-frame to a y-frame by a linear transformation

$$\bar{x}^i = b_j^i y^j \tag{39.12}$$

where the b_j^i are constants. Then $d\bar{x}^i = b_j^i\, dy^j$ and it is a well known result from algebra that the quadratic form $\bar{g}_{ij}d\bar{x}^i d\bar{x}^j$ can be reduced to the diagonal form

$$(dy^1)^2 + (dy^2)^2 + \ldots + (dy^N)^2 \tag{39.13}$$

at the point A by proper choice of the b_j^i (some of these coefficients may have to be given imaginary values). Let h_{ij} be the metric tensor in the y-frame. Then

$$h_{ij} = \frac{\partial \bar{x}^r}{\partial y^i} \frac{\partial \bar{x}^s}{\partial y^j} \bar{g}_{rs} = b_i^r b_j^s \bar{g}_{rs} \tag{39.14}$$

and, in particular, $h_{ij} = \delta_{ij}$ at A. Differentiating the last equation, we obtain

$$\frac{\partial h_{ij}}{\partial y^k} = b_i^r b_j^s b_k^t \frac{\partial \bar{g}_{rs}}{\partial \bar{x}^t} \tag{39.15}$$

showing that $\partial h_{ij}/\partial y^k$ vanishes at A. Thus, in the y-frame, h_{ij} is stationary with value δ_{ij} at A. The implication is that, in a small neighbourhood of A, the coordinates y^i will behave like rectangular Cartesian coordinates. The y-frame is the closest approximation to a rectangular Cartesian frame that can be fitted to the $\mathcal{R}_N$ in the neighbourhood of A.

If the y-frame were exactly rectangular Cartesian (i.e. the space were Euclidean), in order that the law for parallel displacement of a vector, equation (33.7), should agree with the one usually adopted, it would be necessary to take all the components of the affinity to be zero in this frame. It is natural, therefore, to define the affinity at the point A of $\mathcal{R}_N$ so that, in the y-frame, its components are zero. With this choice of affinity, in the y-frame covariant derivatives will reduce to ordinary partial derivatives at the point A; in particular,

$$h_{ij;k} = \frac{\partial h_{ij}}{\partial y^k} = 0 \tag{39.16}$$

at A. But this equation is a tensor equation and, being valid in one frame, is accordingly valid in all frames. Thus, with this choice of affinity

$$g_{ij;k} = 0 \tag{39.17}$$

Assuming that the affinity is defined at all points of $\mathscr{R}_N$ in this manner, the last equation will be valid throughout the space and in all frames. Since this affinity is clearly symmetric in the y-frame, it will be symmetric in all frames.

Writing equation (39.17) out at length, we have

$$\frac{\partial g_{ij}}{\partial x^k} - \Gamma^r_{ik} g_{rj} - \Gamma^r_{jk} g_{ir} = 0 \tag{39.18}$$

Cyclically permuting the indices i, j, k to obtain two further equations, it now follows as from equations (39.7)–(39.9) that

$$[ij, k] = g_{kr} \Gamma^r_{ij} \tag{39.19}$$

Raising the index k, this gives

$$\Gamma^k_{ij} = \{{}_i{}^k{}_j\} \tag{39.20}$$

The affinity determined by this equation in any frame will be called the *metric affinity* and will invariably be assumed in all later developments.

Since equation (39.17) is valid using the metric affinity, the metric tensor behaves like a constant with respect to covariant differentiation. Further, since

$$g^{ij} g_{kj} = \delta^i_k \tag{39.21}$$

by taking the covariant derivative of both members of this equation, we obtain

$$g^{ij}{}_{;m} g_{kj} = 0 \tag{39.22}$$

Multiplying by g^{kr} and summing with respect to k, we then find that

$$g^{ir}{}_{;m} = 0 \tag{39.23}$$

Thus, all forms of the metric tensor behave like constants under covariant differentiation.

It is now clear that

$$(g_{ij} A^j)_{;k} = g_{ij} A^i{}_{;k} \tag{39.24}$$

This shows that the lowering of an index followed by a covariant differentiation yields the same result as when these two processes are reversed. Similarly, it can be shown that the raising of an index and covariant differentiation are two processes which commute.

40. Bianchi identity

If we choose a frame which is geodesic at a point x^i, $\partial g_{ij}/\partial x^k$ will vanish at the point and the two Christoffel symbols will therefore also vanish at the point. Thus, all the components of the affinity will vanish at the point and covariant derivatives will reduce to partial derivatives there. It then follows from equation (39.23) that the partial derivatives $\partial g^{ij}/\partial x^k$ all vanish at the point, also.

In such a geodesic frame, therefore,

$$B^i_{jkl;m} = \frac{\partial}{\partial x^m}\left(\Gamma^i_{rk}\Gamma^r_{jl} - \Gamma^i_{rl}\Gamma^r_{jk} + \frac{\partial \Gamma^i_{jl}}{\partial x^k} - \frac{\partial \Gamma^i_{jk}}{\partial x^l}\right)$$

$$= \frac{\partial^2 \Gamma^i_{jl}}{\partial x^m \partial x^k} - \frac{\partial^2 \Gamma^i_{jk}}{\partial x^m \partial x^l} \tag{40.1}$$

since the Γ^i_{jk} (but not their derivatives necessarily) all vanish at this point. Cyclically permuting the indices k, l, m in equation (40.1), we obtain

$$B^i_{jlm;k} = \frac{\partial^2 \Gamma^i_{jm}}{\partial x^k \partial x^l} - \frac{\partial^2 \Gamma^i_{jl}}{\partial x^k \partial x^m} \tag{40.2}$$

$$B^i_{jmk;l} = \frac{\partial^2 \Gamma^i_{jk}}{\partial x^l \partial x^m} - \frac{\partial^2 \Gamma^i_{jm}}{\partial x^l \partial x^k} \tag{40.3}$$

Addition of equations (40.1), (40.2), (40.3), yields the following identity

$$B^i_{jkl;m} + B^i_{jlm;k} + B^i_{jmk;l} = 0 \tag{40.4}$$

But this is a tensor equation and, having been proved true in the geodesic frame, must be true in all frames. Also, since the chosen point can be any point of $\mathscr{R}_N$, it is valid at all points of the space. It is the *Bianchi identity*.

41. The covariant curvature tensor

The components of B^i_{jkl} are not all independent since the tensor is skew-symmetric in the indices k, l. In addition, however, *if the affinity is symmetric*, it is easily verified from equation (36.21) that

$$B^i_{jkl} + B^i_{klj} + B^i_{ljk} = 0 \tag{41.1}$$

If the affinity is metrical, by lowering the contravariant index of the Riemann–Christoffel tensor, a completely covariant curvature tensor B_{ijkl} is derived. This has a number of symmetry properties, one of which is obtained from our last equation immediately by lowering the index i throughout to give

$$B_{ijkl} + B_{iklj} + B_{iljk} = 0 \tag{41.2}$$

Further such properties can be established by first calculating an expression for the tensor in a geodesic frame. Thus, in such a frame

$$B_{ijkl} = g_{ir}B^r_{jkl} = g_{ir}\left(\frac{\partial \Gamma^r_{jl}}{\partial x^k} - \frac{\partial \Gamma^r_{jk}}{\partial x^l}\right)$$

$$= g_{ir}\frac{\partial}{\partial x^k}(g^{rs}[jl,s]) - g_{ir}\frac{\partial}{\partial x^l}(g^{rs}[jk,s])$$

$$= g_{ir}g^{rs}\left(\frac{\partial}{\partial x^k}[jl,s] - \frac{\partial}{\partial x^l}[jk,s]\right) \tag{41.3}$$

since $\partial g^{rs}/\partial x^k = 0$ in a geodesic frame. Using the result $g_{ir}g^{rs} = \delta^s_i$ and substituting for the Christoffel symbols, we now find from equation (41.3) that

$$B_{ijkl} = \frac{1}{2}\left(\frac{\partial^2 g_{il}}{\partial x^j \partial x^k} + \frac{\partial^2 g_{jk}}{\partial x^i \partial x^l} - \frac{\partial^2 g_{ik}}{\partial x^j \partial x^l} - \frac{\partial^2 g_{jl}}{\partial x^i \partial x^k}\right) \tag{41.4}$$

The following equations are now easily verified:

$$B_{ijkl} = -B_{jikl} \tag{41.5}$$

$$B_{ijkl} = -B_{ijlk} \tag{41.6}$$

$$B_{ijkl} = B_{klij} \tag{41.7}$$

Being true in the geodesic frame, these tensor equations must be valid in all frames. Note that the tensor is skew-symmetric with respect to its first pair of indices and its last pair.

Also lowering the superscript i throughout the Bianchi identity (40.4), we obtain

$$B_{ijkl;m} + B_{ijlm;k} + B_{ijmk;l} = 0 \tag{41.8}$$

42. Divergence. The Laplacian. Einstein's tensor

If the covariant derivative of a tensor field is found and then contracted with respect to the index of differentiation and any superscript, the result is called a *divergence* of the tensor. With respect to orthogonal coordinate transformations in $\mathscr{E}_N$, the partial and covariant derivatives are identical and then this definition of divergence agrees with that given in section 12.

From the tensor A^{ij}_k, two divergences can be formed, viz.

$$\text{div}_i A^{ij}_k = A^{ij}_{k;i} \quad \text{and} \quad \text{div}_j A^{ij}_k = A^{ij}_{k;j} \tag{42.1}$$

A contravariant vector possesses one divergence only, which is an invariant. If the affinity is the metrical one, such a divergence is simply expressed in terms of ordinary partial derivatives thus: since a derivative of a determinant can be found by differentiating each row separately and summing the results, we deduce that

$$\frac{\partial g}{\partial x^j} = G^{ik}\frac{\partial g_{ik}}{\partial x^j} = gg^{ik}\frac{\partial g_{ik}}{\partial x^j} \tag{42.2}$$

Since
$$\frac{\partial g_{ik}}{\partial x^j} = [ij, k] + [kj, i] \tag{42.3}$$

equation (42.2) reduces to

$$\frac{\partial g}{\partial x^j} = g g^{ik}([ij, k] + [kj, i]) = 2g\{_i{}^i{}_j\} \tag{42.4}$$

Hence
$$\{_i{}^i{}_j\} = \frac{1}{\sqrt{g}} \frac{\partial}{\partial x^j}(\sqrt{g}) \tag{42.5}$$

Now let A^i be a vector field. Its divergence is

$$A^i_{;i} = \frac{\partial A^i}{\partial x^i} + \{_{ji}{}^i\} A^j$$

$$= \frac{1}{\sqrt{g}}\left[\sqrt{g}\frac{\partial A^i}{\partial x^i} + A^j \frac{\partial}{\partial x^j}(\sqrt{g}) \right]$$

$$= \frac{1}{\sqrt{g}} \frac{\partial}{\partial x^i}(\sqrt{g} A^i) \tag{42.6}$$

which is the expression required.

In particular, if the vector field is obtained from an invariant V by taking its gradient, we have

$$A_i = \frac{\partial V}{\partial x^i} \tag{42.7}$$

and hence
$$A^i = g^{ij}\frac{\partial V}{\partial x^j} \tag{42.8}$$

From equation (42.6), it now follows that the divergence of this vector is

$$\text{div grad } V = \nabla^2 V = \frac{1}{\sqrt{g}} \frac{\partial}{\partial x^i}\left(\sqrt{g} g^{ij}\frac{\partial V}{\partial x^j} \right) \tag{42.9}$$

The right-hand member of this equation represents the form taken by the *Laplacian* of V in a general Riemannian space. In $\mathscr{E}_N$, employing rectangular axes, $g^{ij} = \delta^{ij}$, $g = 1$ and thus

$$\nabla^2 V = \frac{\partial^2 V}{\partial x^i \partial x^i} \tag{42.10}$$

which is its familiar form.

We shall now calculate the divergence of the Ricci tensor R_{jk} (equation (36.23)). If the metric affinity is being employed, this tensor is symmetric, for

$$R_{kj} = B^i_{kji} = g^{ir}B_{rkji} = g^{ir}B_{jirk} = g^{ir}B_{ijkr}$$
$$= B^r_{jkr} = R_{jk} \tag{42.11}$$

having employed equations (41.5)–(41.7). Raising either index accordingly yields the same mixed tensor R^i_k. If this is contracted, an invariant

$$R = R^j_j \tag{42.12}$$

is obtained. R is called the *curvature scalar* of $\mathscr{R}_N$.

The skew-symmetry of B_{ijkl} with respect to its first two and its last two indices means that equation (41.8) can be writen

$$B_{ijkl;m} - B_{ijml;k} - B_{jimk;l} = 0 \tag{42.13}$$

Multiplying through by $g^{il}g^{jk}$, we then get

$$g^{jk}B^l_{jkl;m} - g^{jk}B^l_{jml;k} - g^{il}B^k_{imk;l} = 0 \tag{42.14}$$

and this is equivalent to

$$g^{jk}R_{jk;m} - g^{jk}R_{jm;k} - g^{il}R_{im;l} = 0 \tag{42.15}$$

or

$$R_{;m} - 2R^k_{m;k} = 0 \tag{42.16}$$

Thus

$$R^k_{m;k} = \tfrac{1}{2}\partial R/\partial x^m \tag{42.17}$$

is the divergence of the Ricci tensor.

Consider now the mixed tensor

$$R^i_j - \tfrac{1}{2}\delta^i_j R \tag{42.18}$$

Its divergence is

$$R^i_{j;i} - \tfrac{1}{2}\delta^i_j \frac{\partial R}{\partial x^i}$$

$$= R^i_{j;i} - \tfrac{1}{2}\frac{\partial R}{\partial x^j}$$

$$= 0 \tag{42.19}$$

This is *Einstein's Tensor*. Its covariant and contravariant components are

$$R_{ij} - \tfrac{1}{2}g_{ij}R, \quad R^{ij} - \tfrac{1}{2}g^{ij}R \tag{42.20}$$

respectively.

43. Geodesics

Let C be any curve constructed in a space $\mathscr{R}_N$ having metric (37.1) and let s be a parameter defined on C such that, if s, $s + ds$ are its values at the respective neighbouring points P, P' on C, then ds is the interval between these two points. If x^i are the coordinates of any point P on C, then the curve will be defined by parametric equations

$$x^i = x^i(s) \tag{43.1}$$

Since dx^i are the components of a vector and ds is an invariant, dx^i/ds is a contravariant vector at P. Its magnitude is, by equation (38.1),

$$\left(g_{ij}\frac{dx^i}{ds}\frac{dx^j}{ds}\right)^{1/2} \tag{43.2}$$

and this is unity by equation (37.1). dx^i/ds is termed the *unit tangent* to the curve at P, its direction being that of the displacement dx^i along the curve from P.

Suppose C possesses the property that the tangents at all its points are parallel, i.e. the curve's direction is constant over its whole length. This property is clearly quite independent of the coordinate frame being employed. In $\mathscr{E}_3$, such a curve would, of course, be a straight line. In $\mathscr{R}_N$, the curve will be called a *geodesic*. A geodesic is accordingly the counterpart of the Euclidean straight line in a Riemannian space. Suppose P, P' are neighbouring points on a geodesic having coordinates $x^i, x^i + dx^i$ respectively. If the unit tangent at P is parallel displaced to P', it will then be identical with the actual unit tangent at this point. Now, by equation (35.4), after parallel displacement from P to P', the unit tangent has components

$$\frac{dx^i}{ds} + \delta\left(\frac{dx^i}{ds}\right) = \frac{dx^i}{ds} - \Gamma^i_{jk}\frac{dx^j}{ds}dx^k \tag{43.3}$$

But the actual unit tangent for the point P' has components

$$\left(\frac{dx^i}{ds}\right)_{s+ds} = \frac{dx^i}{ds} + \frac{d^2x^i}{ds^2}ds \tag{43.4}$$

The vectors (43.3) and (43.4) are identical provided

$$\frac{d^2x^i}{ds^2} + \Gamma^i_{jk}\frac{dx^j}{ds}\frac{dx^k}{ds} = 0 \tag{43.5}$$

If these equations are satisfied at every point of the curve (43.1), it is a geodesic.

The N equations (43.5) are second-order differential equations for the functions $x^i(s)$ and their solution will involve $2N$ arbitrary constants. If A, B are two given points having coordinates $x^i = a^i$, $x^i = b^i$ respectively, the $2N$ conditions that the geodesic must contain these points will, in general, determine the arbitrary constants. Hence there is, in general, a unique geodesic connecting every pair of points. However, in some cases, this will not be so. For example, the geodesics on the surface of a sphere ($\mathscr{R}_2$) are great circles and, in general, there are two great circle arcs joining two given points, a major arc and a minor arc. Also, if these points are diamterically opposed to one another, there is an infinity of great-circle arcs connecting them.

Since dx^i/ds is everywhere a *unit* vector, on a geodesic

$$g_{ij}\frac{dx^i}{ds}\frac{dx^j}{ds} = 1 \tag{43.6}$$

This must, accordingly, be a first integral of the equations (43.5). To show that this is the case, multiply equations (43.5) through by $2g_{ir}dx^r/ds$ and sum with respect to i to obtain

$$2g_{ir}\frac{dx^r}{ds}\frac{d^2x^i}{ds^2} + 2g_{ir}\Gamma^i_{jk}\frac{dx^j}{ds}\frac{dx^k}{ds}\frac{dx^r}{ds} = 0 \tag{43.7}$$

Now
$$2g_{ir}\frac{dx^r}{ds}\frac{d^2x^i}{ds^2} = \frac{d}{ds}\left(g_{ir}\frac{dx^i}{ds}\frac{dx^r}{ds}\right) - \frac{dg_{ir}}{ds}\frac{dx^i}{ds}\frac{dx^r}{ds} \tag{43.8}$$

Also
$$2g_{ir}\Gamma^i_{jk}\frac{dx^j}{ds}\frac{dx^k}{ds}\frac{dx^r}{ds} = 2[jk,r]\frac{dx^j}{ds}\frac{dx^k}{ds}\frac{dx^r}{ds}$$

$$= ([jk,r]+[rk,j])\frac{dx^j}{ds}\frac{dx^k}{ds}\frac{dx^r}{ds}$$

$$= \frac{\partial g_{jr}}{\partial x^k}\frac{dx^k}{ds}\frac{dx^j}{ds}\frac{dx^r}{ds}$$

$$= \frac{dg_{jr}}{ds}\frac{dx^j}{ds}\frac{dx^r}{ds} \tag{43.9}$$

By addition of equations (43.8) and (43.9), it will be seen that equation (43.7) can be expressed in the form

$$\frac{d}{ds}\left(g_{ij}\frac{dx^i}{ds}\frac{dx^j}{ds}\right) = 0 \tag{43.10}$$

Upon integration, there results the first integral

$$g_{ij}\frac{dx^i}{ds}\frac{dx^j}{ds} = \text{constant} \tag{43.11}$$

The constant of integration must, of course, be taken to be unity.

The definition of a geodesic which has been given at the beginning of this section cannot be applied to the class of curves for which the interval ds between adjacent points vanishes. For such a curve, the parametric representation (43.1) is not appropriate and a unit tangent cannot be defined. Instead, suppose that a (1–1) correspondence is set up between the points of the curve and the values of an invariant λ in some interval $\lambda_0 \leqslant \lambda \leqslant \lambda_1$, so that parametric equations for the curve can be written

$$x^i = x^i(\lambda) \tag{43.12}$$

It will be assumed that the derivatives $dx^i/d\lambda$ all exist at each point of the curve. These derivatives constitute a contravariant vector and this has zero magnitude for, since $ds = 0$ along the curve,

$$g_{ij}\frac{dx^i}{d\lambda}\frac{dx^j}{d\lambda} = 0 \tag{43.13}$$

This vector will be in the direction of the displacement vector along the curve dx^i and will be called a *zero tangent* to the curve. The curve will be termed a *null geodesic* if the zero tangents at all points of the curve are parallel. This implies that, when the zero tangent at P is parallel displaced to the adjacent point P′, it must be parallel to the zero tangent at this latter point, and since the magnitudes of these two vectors at P′ are the same, they will be taken to be identical. The condition for this to be so is found, as before, to be

$$\frac{d^2 x^i}{d\lambda^2} + \Gamma^i_{jk} \frac{dx^j}{d\lambda} \frac{dx^k}{d\lambda} = 0 \qquad (43.14)$$

These are, therefore, the equations of the null geodesics. It may now be shown, by an argument similar to that culminating in equation (43.11), that a first integral of these equations is

$$g_{ij} \frac{dx^i}{d\lambda} \frac{dx^j}{d\lambda} = \text{constant} \qquad (43.15)$$

In this case the constant must be zero.

Equation (43.5) may be put in an alternative form which is more convenient for particular calculations, as follows: Multiply through by $2g_{ri}$ and sum with respect to i; the resulting equation is equivalent to

$$\frac{d}{ds}\left(2g_{ri} \frac{dx^i}{ds}\right) - 2\frac{dg_{ri}}{ds}\frac{dx^i}{ds} + 2g_{ri}\Gamma^i_{jk}\frac{dx^j}{ds}\frac{dx^k}{ds} = 0 \qquad (43.16)$$

Now

$$2\frac{dg_{ri}}{ds}\frac{dx^i}{ds} = 2\frac{\partial g_{ri}}{\partial x^k}\frac{dx^k}{ds}\frac{dx^i}{ds}$$

$$= \left(\frac{\partial g_{rj}}{\partial x^k} + \frac{\partial g_{rk}}{\partial x^j}\right)\frac{dx^j}{ds}\frac{dx^k}{ds} \qquad (43.17)$$

and

$$2g_{ri}\Gamma^i_{jk} = [jk, r] = \frac{\partial g_{rj}}{\partial x^k} + \frac{\partial g_{rk}}{\partial x^j} - \frac{\partial g_{jk}}{\partial x^r} \qquad (43.18)$$

Equation (43.16) accordingly reduces to

$$\frac{d}{ds}\left(2g_{ri}\frac{dx^i}{ds}\right) - \frac{\partial g_{jk}}{\partial x^r}\frac{dx^j}{ds}\frac{dx^k}{ds} = 0 \qquad (43.19)$$

Equation (43.14) for a null geodesic may be expressed similarly.

Exercises 5

1. A_{ij} is a covariant tensor. If $B_{ij} = A_{ji}$, prove that B_{ij} is a covariant tensor. Deduce that, if A_{ij} is symmetric (or skew-symmetric) in one frame, it is symmetric (or skew-symmetric) in all. (*Hint*: The equations $A_{ij} = A_{ji}$, $A_{ij} = -A_{ji}$ are tensor equations.)

2. (x, y, z) are rectangular Cartesian coordinates of a point P in $\mathscr{E}_3$ and (r, θ, ϕ) are the corresponding spherical polars related to the Cartesians by equations (30.3). A is a contravariant vector defined at P having components (A^x, A^y, A^z) in

the Cartesian frame and components (A^r, A^θ, A^ϕ) in the spherical polar frame. Express the polar components in terms of the Cartesian components. O1, O2, O3 are rectangular Cartesian axes such that P lies on O1 and O3 lies in the plane Oxy. If (A^1, A^2, A^3) are the components of **A** in this Cartesian frame, show that

$$A^1 = A^r, \quad A^2 = rA^\theta, \quad A^3 = r \sin\theta \, A^\phi$$

(*Note*: Assume the Cartesian axes are right-handed.)

3. If A_i is a covariant vector, verify that $B_{ij} = A_{i,j} - A_{j,i}$ transforms like a covariant tensor. (This is curl **A**.) If **A** is the gradient of a scalar, verify that its curl vanishes.

4. If A_{ij} is a skew-symmetric covariant tensor, verify that

$$B_{ijk} = A_{ij,k} + A_{jk,i} + A_{ki,j}$$

transforms as a tensor.

5. Assuming the transformation inverse to (31.1) exists, prove that each determinant $|\partial \bar{x}^i / \partial x^j|$, $|\partial x^i / \partial \bar{x}^j|$, is the reciprocal of the other. A_j^i is a mixed tensor with respect to this transformation. Show that the determinant $|A_j^i|$ is an invariant.

6. If Γ_{ij}^k is an affinity, show that the torsion defined by

$$T_{ij}^k = \tfrac{1}{2}(\Gamma_{ij}^k - \Gamma_{ji}^k)$$

is a tensor. g_{ij} is a symmetric tensor. Write down an expression for its covariant derivative $g_{ij;k}$. By considering this equation and two similar equations obtained by cyclic permutation of the indices i, j, k, show that if the covariant derivative of g_{ij} is to vanish identically, the affinity must be given by

$$\Gamma_{jk}^i = \{ _j{}^i{}_k \} + T_{jk}^i + g^{ir}(T_{jr}^s g_{sk} + T_{kr}^s g_{sj})$$

7. Show that

$$A_{i;j} - A_{j;i} = A_{i,j} - A_{j,i}$$

provided the affinity is symmetric.

8. Show that

$$A_{i;jk} - A_{i;kj} = B_{ijk}^r A_r + (\Gamma_{kj}^r - \Gamma_{jk}^r) A_{i;r}$$

and deduce that B_{ijk}^r is a tensor and that covariant differentiations are commutative in a space for which $B_{ijk}^r = 0$ and the affinity is symmetric. Obtain the corresponding result for a contravariant vector A^i.

9. A_i is defined at the point x^i and is parallel displaced around a small contour enclosing the point. Prove that the increment in A_i resulting from one circuit is given by

$$\Delta A_i = -\tfrac{1}{2} B_{ijk}^l A_l \alpha^{jk}$$

where α^{jk} is defined by equation (36.10).

10. The parametric equations of a curve in $\mathscr{S}_N$ are

$$x^i = x^i(t)$$

t is an invariant parameter. A tensor A^i_j is defined over a region containing the curve. P, P' are neighbouring points $t, t + \Delta t$ on the curve and ΔA^i_j is defined to be the difference between the actual value of the tensor at P' and the value of the tensor at P after it has been parallel displaced to P'. Prove that

$$\frac{D A^i_j}{Dt} = \lim_{\Delta t \to 0} \frac{\Delta A^i_j}{\Delta t} = A^i_{j;k} \frac{dx^k}{dt}$$

($D A^i_j / Dt$ is called the *intrinsic derivative* of the tensor along the curve.)

11. Verify that $\{^i_{jk}\}$ transforms as an affinity.

12. If A_{ij} is symmetric, prove that $A_{ij,k}$ is symmetric in i and j.

13. Show that the number of the components of B^i_{jkl} which may be assigned values arbitrarily is, in general, $\frac{1}{2} N^3 (N - 1)$. If the affinity is symmetric, show that this number is $\frac{1}{3} N^2 (N^2 - 1)$. (*Hint:* Use equation (41.1).)

14. Show that the number of the components of B_{ijkl} which may be assigned values arbitrarily is $N^2 (N^2 - 1)/12$. (*Hint:* Use equations (41.2), (41.5), (41.6), (41.7).)

15. By differentiating the equation

$$g^{ij} g_{jk} = \delta^i_k$$

with respect to x^l, show that

$$\frac{\partial g^{im}}{\partial x^l} = -g^{mk} g^{ij} \frac{\partial g_{jk}}{\partial x^l}$$

and hence that

$$\frac{\partial g^{im}}{\partial x^l} + g^{ij} \{^m_{jl}\} + g^{mj} \{^i_{jl}\} = 0$$

Deduce that $g^{ij}_{;k} = 0$.

16. If the affinity is the metric one, prove that

$$R_{jk} = B^i_{jki} = -\frac{\partial}{\partial x^i} \{^i_{jk}\} + \frac{\partial^2}{\partial x^j \partial x^k} \log \sqrt{g} + \{^r_{rk}\}\{^i_{ji}\} - \{^r_{jk}\} \frac{\partial}{\partial x^r} \log \sqrt{g}$$

$$S_{kl} = B^i_{ikl} = 0$$

(*Hint:* Employ equation (42.5).) Deduce that R_{jk} is symmetric.

17. If θ, ϕ are co-latitude and longitude respectively on the surface of a sphere of unit radius, obtain the metric

$$ds^2 = d\theta^2 + \sin^2 \theta \, d\phi^2$$

for the surface. Show that the only non-vanishing three index symbols for this $\mathcal{R}_2$ are

$$\{^1_{22}\} = -\sin \theta \cos \theta, \quad \{^2_{12}\} = \{^2_{21}\} = \cot \theta$$

Show also that the only non-vanishing components of B_{ijkl} are

$$B_{1212} = -B_{1221} = B_{2121} = -B_{2112} = \sin^2 \theta$$

and that the components of the Ricci tensor are given by

$$R_{12} = R_{21} = 0, \quad R_{11} = -1, \quad R_{22} = -\sin^2\theta.$$

Prove that the curvature scalar is given by $R = -2$.

18. Employing equation (42.9), obtain expressions for $\nabla^2 V$ in cylindrical and spherical polars.

19. In a certain coordinate system

$$\Gamma^i_{jk} = \delta^i_j \frac{\partial\phi}{\partial x^k} + \delta^i_k \frac{\partial\psi}{\partial x^j}$$

where ϕ, ψ are functions of position. Prove that B^i_{jkl} is a function of ψ only. If $\psi = -\log(a_i x^i)$ prove that

$$R_{jk} = B^i_{jki} = 0$$

20. In the $\mathscr{R}_2$ whose metric is

$$ds^2 = \frac{dr^2 + r^2 d\theta^2}{r^2 - a^2} - \frac{r^2 dr^2}{(r^2 - a^2)^2} \quad (r > a)$$

prove that the differential equation of the geodesics may be written

$$a^2\left(\frac{dr}{d\theta}\right)^2 + a^2 r^2 = k^2 r^4$$

where k^2 is a constant such that $k^2 = 1$ if, and only if, the geodesic is null. By putting $r\, d\theta/dr = \tan\phi$, show that if the space is mapped on a Euclidean plane in which r, θ are taken as polar coordinates, the geodesics are mapped as straight lines, the null geodesics being tangents to the circle $r = a$.

21. A 2-space has metric

$$ds^2 = g_{11}(dx^1)^2 + g_{22}(dx^2)^2$$

where g_{11}, g_{22} are functions of x^1 and x^2. B_{ijkl} is its covariant curvature tensor and R_{ij} is its Ricci tensor. Prove that

$$R_{12} = 0, \quad R_{11}g_{22} = R_{22}g_{11} = B_{1221}$$

If $R = g^{ij}R_{ij}$, show that $R = 2B_{1221}/(g_{11}g_{22})$. Deduce that $R_{ij} = \frac{1}{2} R g_{ij}$.

22. Prove that

(i) $$A^{ij}_{\ ;i} = \frac{1}{\sqrt{g}} \frac{\partial}{\partial x^i}(\sqrt{g} A^{ij}) + A^{ik}\{^{\ j}_{i\ k}\}$$

(ii) $$X^{ij}_{\ ;ij} = 0$$

provided X^{ij} is skew-symmetric. Hence prove that, for any tensor A^{ij}

$$A^{ij}_{\ ;ij} = A^{ij}_{\ ;ji}$$

23. A curve C has parametric equations

$$x^i = x^i(t)$$

and joins two points A and B. The length of the curve is defined to be

$$L = \int_A^B ds = \int_A^B \sqrt{\left(g_{ij}\frac{dx^i}{dt}\frac{dx^j}{dt}\right)}dt$$

Write down the Euler conditions that L should be stationary with respect to all small variations from C and by changing the independent variable in these conditions from t to s, show that they are identical with equations (43.19). (This provides an alternative definition for a geodesic.)

24. If Γ^i_{jk} is a symmetric affinity, show that

$$\Gamma^{i*}_{jk} = \Gamma^i_{jk} + \delta^i_j A_k + \delta^i_k A_j$$

is also a symmetric affinity.

If B^i_{jkl}, B^{i*}_{jkl} are the Riemann–Christoffel curvature tensors relative to the affinities Γ^i_{jk}, Γ^{i*}_{jk} respectively, prove that

$$B^{i*}_{jkl} = B^i_{jkl} + \delta^i_k A_{jl} - \delta^i_l A_{jk} + \delta^i_j(A_{kl} - A_{lk})$$

where $A_{ij} = A_i A_j - A_{i:j}$.

Hence show that if A_i is the gradient of a scalar, then

$$B^{i*}_{jil} - B^{i*}_{lij} = B^i_{jil} - B^i_{lij}$$

25. Prove that the affinity transformations form a group.

26. Prove that

$$\frac{\partial}{\partial x^k}(\nabla\phi)^2 = 2g^{ij}\frac{\partial\phi}{\partial x^i}\phi_{:jk}$$

27. Two metrics are defined in $\mathscr{R}_N$, viz.

$$ds^2 = g_{ij}dx^i dx^j, \quad d\bar{s}^2 = e^\sigma g_{ij}dx^i dx^j$$

where σ is a function of the x^i. If Γ^i_{jk}, $\bar{\Gamma}^i_{jk}$ are metric affinities constructed from these metrics, prove that

$$\bar{\Gamma}^i_{jk} = \Gamma^i_{jk} + A^i_{jk}$$

where

$$A^i_{jk} = \tfrac{1}{2}(\delta^i_j \sigma_{,k} + \delta^i_k \sigma_{,j} - g_{jk}g^{ir}\sigma_{,r})$$

Curvature tensors B^i_{jkl}, $\bar{B}^i_{jkl}$ are constructed from these affinities. Prove that

$$\bar{B}^i_{jkl} = B^i_{jkl} + A^i_{jl;k} - A^i_{jk;l} + A^i_{rk}A^r_{jl} - A^i_{rl}A^r_{jk}$$

Deduce that

$$\overline{R}_{jk} = R_{jk} + A^{i}_{ij;k} - A^{i}_{jk;i} + A^{r}_{rk}A^{i}_{ij} - A^{i}_{ir}A^{r}_{jk}$$

and show also that $A^{i}_{ij} = \frac{1}{2}N\sigma_{,j}$.

28. Oblique Cartesian axes are taken in a plane. Show that the contravariant components of a vector **A** can be obtained by projecting a certain displacement vector on to the axes by parallels to the axes and the covariant components by projecting by perpendiculars to the axes.

29. Define coordinates (r, ϕ) on a right circular cone having semi-vertical angle α so that the metric for the surface is

$$ds^2 = dr^2 + r^2 \sin^2 \alpha \, d\phi^2.$$

Show that the family of geodesics is given by

$$r = a \sec (\phi \sin \alpha - \beta)$$

where a, β are arbitrary constants. Explain this result by developing the cone into a plane.

30. An $\mathscr{R}_N$ has metric

$$ds^2 = e^\lambda \, dx^i \, dx^i$$

where λ is a function of the x^i. Show that the only non-vanishing Christoffel symbols of the second kind are

$$\left\{ \begin{matrix} P \\ Q\,Q \end{matrix} \right\} = (\delta^P_Q - \tfrac{1}{2})\lambda_P, \quad \left\{ \begin{matrix} P \\ P\,Q \end{matrix} \right\} = \tfrac{1}{2}\lambda_Q$$

where $\lambda_r = \partial\lambda/\partial x^r$. Deduce that

$$\left\{ \begin{matrix} i \\ r\,P \end{matrix} \right\} \left\{ \begin{matrix} r \\ P\,i \end{matrix} \right\} = \tfrac{1}{4}(N+2)\lambda^2_P - \tfrac{1}{2}\lambda_r\lambda_r$$

and that the scalar curvature of this space is given by

$$R = (N-1)e^{-\lambda}[\lambda_{rr} + \tfrac{1}{4}(N-2)\lambda_r\lambda_r]$$

where $\lambda_{rr} = \partial^2\lambda/\partial x^r \partial x^r$.

31. θ is the co-latitude and ϕ is the longitude on a unit sphere, so that the metric for the surface is

$$ds^2 = d\theta^2 + \sin^2 \theta \, d\phi^2$$

The covariant vector A_i is taken with initial components (X, Y) and is carried, by parallel displacement, along an arc of length $\phi \sin \alpha$ of the circle $\theta = \alpha$. Show that the components of A_i attain the final values

$$A_1 = X \cos (\phi \cos \alpha) + Y \operatorname{cosec} \alpha \sin (\phi \cos \alpha)$$
$$A_2 = -X \sin \alpha \sin (\phi \cos \alpha) + Y \cos (\phi \cos \alpha)$$

Verify that the magnitude of the vector A_i is unaltered by the displacement.

32. An $\mathscr{R}_3$ has metric

$$ds^2 = \lambda dr^2 + r^2(d\theta^2 + \sin^2\theta\, d\phi^2)$$

where λ is a function of r alone. Show that, along the geodesic for which $\theta = \frac{1}{2}\pi$, $d\theta/ds = 0$ at $s = 0$,

$$\phi = \int \lambda^{1/2}\, d\psi$$

where $r = b\sec\psi$. Interpret this result geometrically when $\lambda = 1$.

33. y^i ($i = 1, 2, 3, 4$) are rectangular Cartesian coordinates in $\mathscr{E}_4$. Show that

$$y^1 = R\cos\theta$$
$$y^2 = R\sin\theta\cos\phi$$
$$y^3 = R\sin\theta\sin\phi\cos\psi$$
$$y^4 = R\sin\theta\sin\phi\sin\psi$$

are parametric equations of a hypersphere of radius R. If (θ, ϕ, ψ) are taken as coordinates on the hypersphere, show that the metric for this $\mathscr{R}_3$ is

$$ds^2 = R^2[d\theta^2 + \sin^2\theta(d\phi^2 + \sin^2\phi\, d\psi^2)]$$

Deduce that in this $\mathscr{R}_3$,

$$B_{1212} = R^2\sin^2\theta, \quad B_{2323} = R^2\sin^4\theta\sin^2\phi,$$
$$B_{3131} = R^2\sin^2\theta\sin^2\phi,$$

all other distinct components being zero. Hence show that

$$B_{ijkl} = K(g_{ik}g_{jl} - g_{il}g_{jk})$$

where $K = 1/R^2$. (This is the condition for the space to be of constant Riemannian curvature K.)

34. An $\mathscr{R}_2$ has metric

$$ds^2 = \text{sech}^2 y(dx^2 + dy^2)$$

Find the equation of the family of geodesics.

θ, ϕ are co-latitude and longitude respectively on the surface of a sphere of unit radius. Mercator's projection is obtained by plotting x, y as rectangular Cartesian coordinates in a plane, taking

$$x = \phi, \quad y = \log\cot\tfrac{1}{2}\theta$$

Calculate the metric for the spherical surface in terms of x and y and deduce that the great circles are represented by the curves

$$\sinh y = \alpha\sin(x + \beta),$$

where α, β are parameters, in Mercator's projection.

35. Obtain the formula

$$B_{ijkl} = \frac{1}{2}\left(\frac{\partial^2 g_{il}}{\partial x^j \partial x^k} + \frac{\partial^2 g_{jk}}{\partial x^i \partial x^l} - \frac{\partial^2 g_{ik}}{\partial x^j \partial x^l} - \frac{\partial^2 g_{jl}}{\partial x^i \partial x^k}\right) + g_{sr}\begin{Bmatrix} r \\ i \ l \end{Bmatrix}\begin{Bmatrix} s \\ j \ k \end{Bmatrix}$$

$$-g_{sr}\begin{Bmatrix} r \\ i \ k \end{Bmatrix}\begin{Bmatrix} s \\ j \ l \end{Bmatrix}$$

36. An $\mathcal{R}_2$ has metric $ds^2 = 2\phi\, dx\, dy$, where $\phi = \phi(x, y)$. Calculate the component B_{1212} of the curvature tensor and state the values of the remaining 15 components. Deduce that the space is flat provided

$$\phi\frac{\partial^2 \phi}{\partial x \partial y} = \frac{\partial \phi}{\partial x}\frac{\partial \phi}{\partial y}$$

Putting $\phi = e^{\psi}$, obtain the general form for ϕ satisfying this condition. Deduce that coordinates ξ, η can be found such that the metric takes the form $ds^2 = 2d\xi\, d\eta$.

37. If A_i is such that $A_{i;j} + A_{j;i} = 0$, by cyclically permuting the indices i, j, k in $A_{i;jk} - A_{i;kj} = A_r B^r_{ijk}$ to give two further equations, prove that $A_{i;jk} = -A_r B^r_{kij}$.

38. $\phi(u, v)$ and $\psi(u, v)$ are the real and imaginary parts of an analytic function $f(w)$ of the complex variable $w = u + iv$. Show that the equations $x = \phi(u, v)$, $y = \psi(u, v)$ transform the Pythagorean metric $ds^2 = dx^2 + dy^2$ into the metric

$$ds^2 = \left[\left(\frac{\partial \phi}{\partial u}\right)^2 + \left(\frac{\partial \phi}{\partial v}\right)^2\right](du^2 + dv^2).$$

By taking $f(w) = 1/w$, explain how it is possible to write down the equation of the family of geodesics in a space whose metric is

$$ds^2 = \frac{du^2 + dv^2}{(u^2 + v^2)^2}$$

Also obtain this equation by transforming the metric using the equations $u = r\cos\theta$, $v = r\sin\theta$, and writing down the differential equations for the geodesics in terms of r and θ.

39. (x, y) are rectangular Cartesian coordinates and (r, θ) are polar coordinates in a Euclidean plane. A_{ij} is a symmetric tensor field defined in the plane by its components $A_{xx} = A_{yy} = 0$, $A_{xy} = A_{yx} = x/y + y/x$. Calculate the contravariant polar components of the field in terms of r and θ, and deduce that $A^{rr} + r^2 A^{\theta\theta} = 0$. (Ans. $A^{rr} = 2$, $A^{r\theta} = 2\cot 2\theta/r$, $A^{\theta\theta} = -2/r^2$.)

40. x, y, z are rectangular Cartesian coordinates in $\mathcal{E}_3$. Parametric equations for a hyperbolic paraboloid are taken in the form $x = u + v$, $y = u - v$, $z = uv$. A covariant tensor field on the surface is defined by the equations $A_{uu} = u^2$, $A_{uv} = A_{vu} = -uv$, $A_{vv} = v^2$. Show that the contravariant components are one-quarter the covariant components.

41. x, y are rectangular Cartesian coordinates in a Euclidean plane and u, v are curvilinear coordinates defined by $x = a \cosh u \cos v$, $y = a \sinh u \sin v$. A covariant vector has components A_x, A_y at the point (x, y) and curvilinear components A_u, A_v. Show that

$$A_x = \frac{2}{a}(A_u \sinh u \cos v - A_v \cosh u \sin v)/(\cosh 2u - \cos 2v)$$

42. x, y are rectangular Cartesian coordiantes in a plane. Curvilinear coordinates u, v are defined by the transformation equations $u = \frac{1}{2}(x^2 - y^2)$, $v = xy$. Sketch the families of coordinate lines $u = $ const., $v = $ const. and show that the metric in the uv-frame is

$$ds^2 = \tfrac{1}{2}(u^2 + v^2)^{-1/2}(du^2 + dv^2)$$

A covariant vector has Cartesian components (A_x, A_y) and curvilinear components (A_u, A_v). Show that

$$A_u = (xA_x - yA_y)/(x^2 + y^2)$$

and derive the corresponding formula for A_v.

43. (x, y) are rectangular Cartesian coordinates in a Euclidean plane and (u, v) are curvilinear coordinates defined by the equations $x = \frac{1}{2}(u^2 + v^2)$, $y = uv$. A covariant vector field A_i is defined over the plane in the uv-frame by the equations $A_u = A_v = (u^2 - v^2)^2$. Show that its divergence is equal to $2(u - v)^2/(u + v)$. Calculate the contravariant vector at the point $u = 0$, $v = 1$ and use the law of parallel displacement along the curve $u = 0$ to calculate the parallel displaced vector at the point $u = 0$, $v = 2$. (Ans. $A^u = A^v = \frac{1}{2}$.)

44. An $\mathcal{R}_2$ has metric $ds^2 = dx^2 + x^2 dy^2$. Calculate the components of its metric affinity. Deduce that the divergence of the vector field whose covariant components are given by $A_x = x \cos 2y$, $A_y = -x^2 \sin 2y$, vanishes.

45. If the x-frame is geodesic at the point x^i, prove that

$$\frac{\partial^2}{\partial x^i \partial x^j}\begin{Bmatrix} r \\ s\ t \end{Bmatrix} = \tfrac{1}{2}g^{rp}(g_{sp,ijt} + g_{tp,ijs} - g_{st,ijp})$$

$$R^i_{j;i} = g^{ik}\left[\frac{\partial^2}{\partial x^i \partial x^k}\begin{Bmatrix} s \\ j\ s \end{Bmatrix} - \frac{\partial^2}{\partial x^i \partial x^s}\begin{Bmatrix} s^! \\ j\ k \end{Bmatrix}\right]$$

$$R_{,j} = g^{ik}\left[\frac{\partial^2}{\partial x^j \partial x^k}\begin{Bmatrix} s \\ i\ s \end{Bmatrix} - \frac{\partial^2}{\partial x^j \partial x^s}\begin{Bmatrix} s \\ i\ k \end{Bmatrix}\right]$$

and deduce that $R^i_{j;i} = \frac{1}{2}R_{,j}$.

46. An $\mathcal{R}_2$ has metric $ds^2 = y^2 dx^2 + dy^2$. Deduce the parallel transfer equations

$$\delta A_x = \frac{1}{y}A_x dy - yA_y dx, \quad \delta A_y = \frac{1}{y}A_x dx$$

Using these equations, parallel transfer the vector along the curve $y = \sec x$ from the point $x = 0$, $y = 1$ at which its components are $A_x = 0$, $A_y = 1$, to the point $x = \pi/3$, $y = 2$.

47. If θ, ϕ are latitude and longitude respectively on the surface of a globe of unit radius, show that the geodesics on the globe have equations $\tan \theta = \tan \alpha \sin (\phi + \beta)$, where α, β are constants.

48. A space $\mathscr{R}_2$ has metric $ds^2 = \operatorname{sech}^2 y (dx^2 + dy^2)$. A vector A_i is parallel displaced from the point $x = 0$, $y = b$ to the point $x = a$, $y = b$ along the line $y = b$. Its initial components are (X, Y). Show that its final components are given by

$$A_1 = X \cos (a \tanh b) + Y \sin (a \tanh b)$$
$$A_2 = - X \sin (a \tanh b) + Y \cos (a \tanh b)$$

49. Replacing A_i in the argument of section 34 by A^i, obtain the transformation law for an affinity in the form

$$\bar{\Gamma}^i_{jk} = \frac{\partial \bar{x}^i}{\partial x^r} \frac{\partial x^s}{\partial \bar{x}^j} \frac{\partial x^t}{\partial \bar{x}^k} \Gamma^r_{st} - \frac{\partial^2 \bar{x}^i}{\partial x^r \partial x^s} \frac{\partial x^r}{\partial \bar{x}^j} \frac{\partial x^s}{\partial \bar{x}^k}$$

Prove that this is equivalent to equation (34.8). (*Hint:* Differentiate $\dfrac{\partial \bar{x}^i}{\partial x^r} \dfrac{\partial x^r}{\partial \bar{x}^j} = \delta^i_j$.)

50. Using the transformation law for an affinity in the form given in the last exercise, if $\Gamma^i = g^{jk} \Gamma^i_{jk}$ show that an $\bar{x}$-frame can always be found such that $\bar{\Gamma}^i = 0$ everywhere. Show that this frame is determined by the equations

$$g^{jk} \frac{\partial^2 \bar{x}^i}{\partial x^j \partial x^k} = \frac{\partial \bar{x}^i}{\partial x^j} \Gamma^j$$

and that these equations can be written $\nabla^2 \bar{x}^i = 0$. (The coordinates $\bar{x}^i$ are said to be *harmonic*.)

CHAPTER 6

General Theory of Relativity

44. Principle of equivalence

The special theory of relativity rejects the Newtonian concept of a privileged observer, at rest in absolute space, and for whom physical laws assume their simplest form, and assumes instead that these laws will be identical for all members of a class of inertial observers in uniform translatory motion relative to one another. Thus, although the existence of a *single* privileged observer is denied, the existence of a *class* of such observers is accepted. This seems to imply that, if all matter in the universe were annihilated except for a single experimenter and his laboratory, this observer would, nonetheless, be able to distinguish inertial frames from non-inertial frames by the special simplicity which the descriptions of physical phenomena take with respect to the former. The further implication is, therefore, that physical space is not simply a mathematical abstraction which it is convenient to employ when considering distance relationships between material bodies, but exists in its own right as a separate entity with sufficient internal structure to permit the definition of inertial frames. However, all the available evidence suggests that physical space cannot be defined except in terms of distance measurements between physical bodies. For example, such a space can be constructed by setting up a rectangular Cartesian coordinate frame comprising three mutually perpendicular rigid rods and then defining the coordinates of the point occupied by a material particle by distance measurements from these rods in the usual way. Physical space is, then, nothing more than the aggregate of all possible coordinate frames. A claim that physical space exists independently of distance measurements between material bodies, can only be substantiated if a precise statement is given of the manner in which its existence can be detected without carrying out such measurements. This has never been done and we shall assume, therefore, that the special properties possessed by inertial frames must be related in some way to the distribution of matter within the universe and that they are not an indication of an inherent structure possessed by physical space when it is considered apart from the matter it contains. This line of argument encourages us to expect, therefore, that, ultimately, all physical laws will be expressible in forms which are quite independent of any coordinate frame by which physical space is defined, i.e. that physical laws are identical for all observers. This is the

127

general principle of relativity. This does not mean that, when account is taken of the actual distribution of matter within the universe, certain frames will not prove to be more convenient than others. When calculating the field due to a distribution of electric charge, it simplifies the calculations enormously if a reference frame can be employed relative to which the charge is wholly at rest. However, this does not mean that the laws of electromagnetism are expressible more simply in this frame, but only that this particular charge distribution is then described more simply. Similarly, we shall attribute the simpler forms taken by some calculations when carried out in inertial frames, to the special relationship these frames bear to the matter present in the universe. Fundamentally, therefore, all observers will be regarded as equivalent and, by employing the same physical laws, will arrive at identical conclusions concerning the development of any physical system.

The main difficulty which arises when we try to express physical laws so that they are valid for all observers is that, if test particles are released and their motions studied from a frame which is being accelerated with respect to an inertial frame, these motions will not be uniform and this fact appears to set such frames apart from inertial frames as a special class for which the ordinary laws of motion do not apply. However, by a well-known device of Newtonian mechanics, viz. the introduction of *inertial forces*, accelerated frames can be treated as though they were inertial and this suggests a way out of our difficulty. Thus suppose a space rocket, moving *in vacuo*, is being accelerated uniformly by the action of its motors. An observer inside the rocket will note that unsupported particles experience an acceleration parallel to the axis of the rocket. Knowing that the motors are operating, he will attribute this acceleration to the fact that his natural reference frame is being accelerated relative to an inertial frame. However he may, if he prefers, treat his reference frame as inertial and suppose that all bodies within the rocket are being subjected to inertial forces acting parallel to the rocket's axis. If **a** is the acceleration of the rocket, the appropriate inertial force to be applied to a particle of mass m is $-m\mathbf{a}$. Similarly, if the rocket's motors are shut down but the rocket is spinning about its axis, an observer within the rocket will again note that free particles do not move uniformly relative to his surroundings and he may again avoid attributing this phenomenon to the fact that his frame is not inertial, by supposing certain inertial forces (viz. centrifugal and Coriolis forces) to act upon the particles. Now it is an obvious property of each such inertial force that it must cause an acceleration which is independent of the mass of the body upon which it acts, for the force is always obtained by multiplying the body's mass by an acceleration independent of the mass. This property it shares with a gravitational force, for this also is proportional to the mass of the particle being attracted and hence induces an acceleration which is independent of this mass. This independence of the gravitational acceleration of a particle and its mass has been checked experimentally with great accuracy by Eötvös. If, therefore, we regard the equivalence of inertial and gravitational forces as having been established, inertial forces can be thought of as arising from the presence of gravitational fields. This is the *principle of equivalence*. By this principle, in the case of the uniformly accelerated rocket, the observer is entitled to neglect his acceleration relative to an inertial frame, provided he accepts the existence of a uniform gravitational field of intensity – **a** parallel to the axis of the rocket. Similarly, the observer in the rotating rocket may disregard his motion and accept, instead, the existence of a gravitational field having such a nature as to account for the centrifugal and Coriolis forces.

By appeal to the principle of equivalence, therefore, an observer employing a reference frame in arbitrary motion with respect to an inertial frame, may disregard this motion and assume, instead, the existence of a gravitational field. The intensity of this field at any point within the frame will be equal to the inertial force per unit mass at the point. By this device, every observer becomes entitled to treat his reference frame as being at rest and all observers accordingly become equivalent. However, the reader is probably still not convinced that the distinction between accelerated and inertial frames has been effectively eliminated, but only that it has been concealed by means of a mathematical device having no physical significance. Thus, he may point out that the gravitational fields which have been introduced to account for the inertial forces are 'fictitious' fields, which may be completely removed by choosing an inertial frame for reference purposes, whereas 'real' fields, such as those due to the earth and sun, cannot be so removed. He may further object that no physical agency can be held responsible for the presence of a 'fictitious' field, whereas a 'real' field is caused by the presence of a massive body. These objections may be met by attributing such 'fictitious' fields to the motions of distant masses within the universe. Thus, if an observer within the uniformly accelerated rocket takes himself to be at rest, he must accept as an observable fact that all bodies within the universe, including the galaxies, possess an additional acceleration of –**a** relative to him and to this motion he will be able to attribute the presence of the uniform gravitational field which is affecting his test particles. Again, the whole universe will be in rotation about the observer who regards himself and his space-ship as stationary when it is in rotation relative to an inertial frame. It is this rotation of the masses of the universe which we shall hold responsible for the Coriolis and centrifugal gravitational fields within the rocket. But, in addition, these 'inertial' gravitational fields will account for the motions of the galaxies as observed from the non-inertial frame. Thus, for the observer within the uniformly accelerated rocket a uniform gravitational field of intensity – **a** extends over the whole of space and is the cause of the acceleration of the galaxies; for the observer within the rotating rocket, the resultant of the centrifugal and Coriolis fields acting upon the galaxies is just sufficient to account for their accelerations in their circular orbits about himself as centre (the reader should verify this, employing the results of Exercises 1, No. 1). On this view, therefore, inertial frames possess particularly simple properties only because of their special relationship to the distribution of mass within the universe. In much the same way, the electromagnetic field due to a distribution of electric charge takes an especially simple form when described relative to a frame in which all the charges are at rest (assuming such exists). If any other frame is employed, the field will be complicated by the presence of a magnetic component arising from the motions of the charges. However, this magnetic field is not considered imaginary because a frame can be found in which it vanishes, whereas for certain magnetic fields such a frame cannot be found. The laws of electromagnetism are taken to be valid in all frames, though it is conceded that, for solving particular problems, a certain frame may prove to be pre-eminently more convenient than any other. Neither, therefore, should the centrifugal and Coriolis fields be dismissed as imaginary solely because they can be removed by proper choice of a reference frame, although it may be convenient to make such a choice of frame when carrying out particular computations. In short, the general principle of relativity can be accepted as valid and, at the same time, the existence of the inertial frames accounted for by the simplicity of the motions of the galactic masses with respect to these particular frames.

The notion that the existence of inertial frames is bound up with the large-scale distribution of matter within the universe is referred to as *Mach's principle*. Although Einstein was powerfully influenced by the principle when developing his general theory, he was disappointed to discover that it still permits the existence of universes in which local inertial frames are not in uniform non-rotatory motion relative to the overall matter distribution. The complete integration of Mach's principle into the theory is yet to be accomplished.

The previously unexplained identity of inertial and gravitational masses is easily deduced as a consequence of the principle of equivalence. For, consider a particle of mass m which is being observed from a non-inertial frame. A gravitational force equal to the inertial force will be observed to act upon this body. This force is directly proportional to the inertial mass m. But, by the principle of equivalence, all gravitational forces are of the same nature as this particular force and will, accordingly, be directly proportional to the inertial masses of the bodies upon which they act. Thus the gravitational 'charge' of a particle, measuring its susceptibility to the influences of gravitational fields, is identical with its inertial mass and the identity of inertial and gravitational masses has been explained in a straightforward and convincing manner.

45. Metric in a gravitational field

Suppose that a space-station in the shape of a wheel has been constructed in a region of space far from other attracting bodies and that it is set rotating in its plane about its centre with angular velocity ω. An observer O, wearing a space-suit, is located outside the station and does not participate in the rotary motion; his frame of reference is therefore inertial. O watches C, a member of the station's crew, measuring the dimensions of the station using a metre rule. C first measures the radius of the station from its centre to its outer wall by laying his rule along one of the corridors forming a spoke of the wheel. O notes that the rule is moving laterally throughout the measuring process, but this motion does not affect its length in his frame and he will accordingly agree with the radius r recorded by C. C next lays his rule around the outer wall of the station and records a perimeter p. During this process, however, O sees the rule moving longitudinally with velocity ωr and its length will be reduced by a factor $\sqrt{(1 - \omega^2 r^2/c^2)}$. He will accordingly correct the length of the perimeter found by C to the value $p \sqrt{(1 - \omega^2 r^2/c^2)}$. Since O's frame is inertial, Euclidean geometry is valid for all space measurements referred to the frame and he must find that

$$p \sqrt{(1 - \omega^2 r^2/c^2)} = 2\pi r \qquad (45.1)$$

Thus $\qquad\qquad p = 2\pi r (1 - \omega^2 r^2/c^2)^{-1/2} \qquad (45.2)$

This last equation indicates that C will discover that the Euclidean formula $p = 2\pi r$ is not valid for measurements made in the rotating frame of the space-station. But C is entitled to regard the station frame as being at rest, provided he accepts the existence of a gravitational field which will account for the centrifugal and Coriolis forces he experiences. We conclude that, relative to a frame at rest in such a gravitational field, spatial measurements will not be in conformity with Euclidean geometry.

By the principle of equivalence, the conclusion which has just been reached concerning the non-Euclidean nature of space in which there is present a gravitational field of the centrifugal–Coriolis type, must be extended to all gravitational fields. However, in the case of a field such as that which surrounds the earth, it will not be possible (as it is for the centrifugal–Coriolis field) to find an inertial frame of reference relative to which the field vanishes and for which the spatial geometry is Euclidean. Such a field will be termed *irreducible*. Even in an irreducible field, however, a frame can always be found which is inertial for a sufficiently small region of space and a sufficiently small time duration. Thus, within a space-ship which is not rotating relative to the extragalactic nebulae and which is falling freely in the earth's gravitational field, free particles will follow straight-line paths at constant speed for considerable periods of time and the conditions will be inertial. A coordinate frame fixed in the ship will accordingly simulate an inertial frame over a restricted region of space and time and its geometry will be approximately Euclidean.

Since a rectangular Cartesian coordinate frame can be set up only in a space possessing a Euclidean metric, this method of specifying the relative positions of events must be abandoned in an irreducible gravitational field (except over small regions as has just been explained). Instead, the positions and times of all events will be specified by reference to a very general type of frame which we can suppose constructed as follows: Imagine the whole of the cosmos is filled by a fluid whose motion is arbitrary but non-turbulent (i.e. particles of the fluid which are initially close together, remain in proximity to one another). Let each molecule of the fluid be a clock which runs smoothly, but not necessarily at a constant rate as judged by a standard atomic clock. No attempt will be made to synchronize clocks which are separated by a finite distance, but it will be assumed that, as this distance tends to zero, the readings of the clocks will always approach one another. Each clock will be allocated three spatial coordinates ξ^1, ξ^2, ξ^3 according to any scheme which ensures that the coordinates of adjacent clocks only differ infinitesimally. The coordinates ξ^α of a clock will be supposed never to change. Any event taking place anywhere in the cosmos can now be allocated unique space–time coordinates ξ^i ($i = 1, 2, 3, 4$) as follows: (ξ^1, ξ^2, ξ^3) are the spatial coordinates belonging to the clock which happens to be adjacent to the event when it occurs, and ξ^4 is the time shown on this clock at this instant.

We shall now further generalize the coordinates allocated to an event. Let x^i ($i = 1, 2, 3, 4$) be any functions of the ξ^i such that, to each set of values of the ξ^i there corresponds one set of values of the x^i, and conversely. We shall write

$$x^i = x^i(\xi^1, \xi^2, \xi^3, \xi^4) \tag{45.3}$$

Then the x^i, also, will be accepted as coordinates, with respect to a new frame of reference, of the event whose coordinates were previously taken to be the ξ^i. It should be noted that, in general, each of the new coordinates x^i will depend upon both the time and the position of the event, i.e., it will not necessarily be the case that three of the coordinates x^i are spatial in nature and one is temporal. All possible events will now be mapped upon a space $\mathscr{S}_4$, so that each event is represented by a point of the space and the x^i will be the coordinates of this point with respect to a coordinate frame. $\mathscr{S}_4$ will be referred to as the *space–time continuum*.

It has been remarked that, in any gravitational field, it is always possible to define a frame relative to which the field vanishes over a restricted region and which behaves as an inertial frame for events occurring in this region and extending over a small interval of time. Such a frame will be falling freely in the gravitational field and will accordingly be referred to as a *local free-fall frame*. Suppose, then, that such an inertial frame S is found for two contiguous events. Any other frame in uniform motion relative to S will also be inertial for these events. Observers at rest in all such frames will be able to construct rectangular Cartesian axes and synchronize their standard atomic clocks in the manner described in Chapter 1 and hence measure the proper time interval $d\tau$ between the events. If, for one such observer, the events at the points having rectangular Cartesian coordinates (x, y, z), $(x + dx, y + dy, z + dz)$ occur at the times $t, t + dt$ respectively, then

$$d\tau^2 = dt^2 - \frac{1}{c^2}(dx^2 + dy^2 + dz^2) \tag{45.4}$$

The *interval* between the events ds will be defined by

$$ds^2 = -c^2 \, d\tau^2 = dx^2 + dy^2 + dz^2 - c^2 \, dt^2 \tag{45.5}$$

The coordinates (x, y, z, t) of an event in this quasi-inertial frame will be related to the coordinates x^i defined earlier, by equations

$$x = x(x^1, x^2, x^3, x^4), \qquad \text{etc.} \tag{45.6}$$

and hence
$$dx = \frac{\partial x}{\partial x^i} dx^i, \quad \text{etc.} \tag{45.7}$$

Substituting for dx, dy, dz, dt in equation (45.5), we obtain the result

$$ds^2 = g_{ij} dx^i dx^j \tag{45.8}$$

determining the interval ds between two events contiguous in space–time, relative to a general coordinate frame valid for the whole of space–time. The space–time continuum can accordingly be treated as a Riemannian space with metric given by equation (45.8).

As explained in section 7, $d\tau$ can be timelike or spacelike according as it is real or imaginary respectively. If it is real, it will be possible for a standard clock to be present at both the events x^i, $x^i + dx^i$ and the time which elapses between them as measured by this clock will be $d\tau$. Alternatively, if ds is imaginary, ds/ic can be interpreted as the time between two contiguous events as measured by a standard clock present at both.

46. Motion of a free particle in a gravitational field

In a region of space which is at a great distance from material bodies, rectangular Cartesian axes $Oxyz$ can be found constituting an inertial frame. If time is measured by clocks synchronized within this frame and moving with it, the motion of a freely moving test particle relative to the frame will be uniform. Thus, if (x, y, z) is the position of such a particle at time t, its equations of motion can be written

$$\frac{d^2x}{dt^2} = \frac{d^2y}{dt^2} = \frac{d^2z}{dt^2} = 0 \tag{46.1}$$

Let ds be the interval between the event of the particle arriving at the point (x, y, z) at time t and the contiguous event of the particle arriving at $(x + dx, y + dy, z + dz)$ at $t + dt$. Then ds is given by equation (45.5) and, if v is the speed of the particle, it follows from this equation that

$$ds = (v^2 - c^2)^{1/2} dt \tag{46.2}$$

Since v is constant, it now follows that equations (46.1) can be expressed in the form

$$\frac{d^2x}{ds^2} = \frac{d^2y}{ds^2} = \frac{d^2z}{ds^2} = 0 \tag{46.3}$$

Also, from equation (46.2) it may be deduced that

$$\frac{d^2t}{ds^2} = 0 \tag{46.4}$$

Equations (46.3) and (46.4) determine the family of world-lines of free particles in space–time relative to an inertial frame.

Now suppose that any other reference frame and procedure for measuring time is adopted in this region of space, e.g. a frame which is in uniform rotation with respect to an inertial frame might be employed. Let (x^1, x^2, x^3, x^4) be the coordinates of an event in this frame. The interval between two contiguous events will then be given by equation (45.8). If an observer using this frame releases a test particle and observes its motion relative to the frame, he will note that it is not uniform or even rectilinear and will be able to account for this fact by assuming the presence of a gravitational field. He will find that the particle's equations of motion are

$$\frac{d^2 x^i}{ds^2} + \left\{ {}^i_{jk} \right\} \frac{dx^j}{ds} \frac{dx^k}{ds} = 0 \tag{46.5}$$

This must be the case for, as shown in section 43, this is a tensor equation defining a geodesic and valid in every frame if it is valid in one. But, in the $xyzt$-frame, the g_{ij} are all constant and the three index symbols vanish. Hence, in this frame, the equations (46.5) reduce to the equations (46.3) and (46.4) and these are known to be true for the particle's motion. We have shown, therefore, that the effect of a gravitational field of the reducible variety upon the motion of a test particle can be allowed for when the form taken by the metric tensor g_{ij} of the space–time manifold is known relative to the frame being employed. This means that the g_{ij} determine, and are determined by, the gravitational field.

The ideas of the previous paragraph will now be extended to regions of space where irreducible gravitational fields are present. It has been pointed out that, for any sufficiently small region of such space and interval of time, an inertial frame can be found and consequently the paths of freely moving particles will be governed in such a small region by equations (46.5). It will now be assumed that these are the equations of motion of free particles without any restriction, i.e. that the world-line of a free particle is a geodesic for the space–time manifold or that the world-line of a free particle has constant direction. This appears to be the natural generalization of the Galilean law of inertia whereby, even in an irreducible gravitational field, a particle's trajectory through space–time is the straightest possible after consideration has been given to the intrinsic curvature of the continuum. It will then follow that the motions of particles falling freely in any gravitational field can be determined relative to any frame when the components g_{ij} of the metric tensor for this frame are known. Thus the g_{ij} will always specify the gravitational field observed to be present in a frame and the only distinction between irreducible and reducible fields will be that, for the latter it will be possible to find a coordinate frame in space–time for which the metric tensor has all its components zero except

$$g_{11} = g_{22} = g_{33} = 1 \quad g_{44} = -c^2 \tag{46.6}$$

whereas for the former this will not be possible.

It will be proved in section 50 that the assumption we are making can be derived from Einstein's law of gravitation and hence does not constitute an additional basic hypothesis of the theory.

Since the Christoffel symbols vanish in a frame which is geodesic at some point of space–time, in such a frame equations (46.5) reduce to $d^2x^i/ds^2 = 0$ over a small neighbourhood of the point. If, in addition, the frame is chosen to be quasi-Euclidean with metric (45.5), equations (46.1) will be valid over the neighbourhood and a freely falling body will have very nearly uniform motion. Such a frame can therefore be identified with a local freely falling frame.

47. Einstein's law of gravitation

According to Newtonian ideas, the gravitational field which exists in any region of space is determined by the distribution of matter. This suggests that the metric tensor of the space–time manifold, which has been shown to be closely related to the observed gravitational field, should be calculable when the matter distribution throughout space–time is known. We first look, therefore, for a tensor quantity describing this matter distribution with respect to any frame in space–time and then attempt to relate this to the metric tensor. The energy–momentum tensor T_{ij}, defined in section 21 with respect to an inertial frame, immediately suggests itself. Both matter and electromagnetic energy contribute to the components of this tensor but since, according to the special theory, mass and energy are basically identical, it is to be expected that all forms of energy, including the electromagnetic variety, will contribute to the gravitational field.

Since the energy–momentum tensor has been defined in inertial frames only, this definition must now be extended to apply to a general coordinate frame in space–time. This can be carried out thus: In the neighbourhood of a point P of space–time, a frame with coordinates y^i can be defined which is geodesic at P and whose metric reduces to the Euclidean form (39.13) at P. As explained in the last section, this frame will correspond to a local freely falling quasi-inertial frame in which the y^i will behave like Minkowski coordinates; we shall assume that the equations of the special theory are valid in this frame at P. The transformation equations relating the coordinates y^i of an event to its coordinates x^i with respect to any other coordinate frame can now be found. Then, if $T_{ij}^{(0)}$ are the components of the energy–momentum tensor in the y-frame at the point P, its components in the x-frame at this point can be determined from the appropriate tensor transformation equations. Thus, the covariant energy–momentum tensor will have components T_{ij} in the x-frame given by

$$T_{ij} = \frac{\partial y^r}{\partial x^i} \frac{\partial y^s}{\partial x^j} T_{rs}^{(0)} \tag{47.1}$$

Since covariant and contravariant tensors are indistinguishable with respect to rectangular Cartesian axes, $T_{rs}^{(0)}$ can also be taken to be the components of a contravariant tensor in the y-frame and the components of this tensor in the x-frame will then be given by the equation

$$T^{ij} = \frac{\partial x^i}{\partial y^r} \frac{\partial x^j}{\partial y^s} T_{rs}^{(0)} \tag{47.2}$$

Similarly, the components of the mixed energy–momentum tensor are given by

$$T_j^i = \frac{\partial x^i}{\partial y^r} \frac{\partial y^s}{\partial x^j} T_{rs}^{(0)} \tag{47.3}$$

These transformations can be carried out at every point of space–time, thus generating for the x-frame an energy–momentum tensor field throughout the continuum. It is left as an exercise for the reader to show that the last three equations are consistent, i.e. raising the indices of T_{ij} as given by equation (47.1) leads to T^{ij} as given by equation (47.2) (see Exercise 1 at the end of this chapter).

Consider the tensor equation

$$T^{ij}_{\;;j} = 0 \tag{47.4}$$

Expressed in terms of the coordinates y^i at any point of space–time, this simplifies to

$$T^{(0)}_{ij,\,j} = 0 \tag{47.5}$$

which is equation (21.20). Being valid in one frame, therefore, equation (47.4) is true for all frames. Thus, the divergence of the energy–momentum tensor vanishes. If, therefore, this tensor is to be related to the metric tensor g_{ij}, the relationship should be of such a form that it implies equation (47.4). Now

$$g^{ij}_{\;;k} = 0 \tag{47.6}$$

by equation (39.23) and hence, *a fortiori*,

$$g^{ij}_{\;;j} = 0 \tag{47.7}$$

The law
$$T^{ij} = \lambda g^{ij} \tag{47.8}$$

where λ is a universal constant, would accordingly be satisfactory in this respect. However, over a region in which matter and energy were absent so that $T^{ij} = 0$, this would imply that

$$g^{ij} = 0 \tag{47.9}$$

which is clearly incorrect. Further, according to Newtonian theory, if μ is the density of matter, the gravitational field can be derived from a potential function U satisfying the equation

$$\nabla^2 U = 4\pi G \mu \tag{47.10}$$

where G is the gravitational constant. The new law of gravitation which is being sought must include equation (47.10) as an approximation. But, as appears from equation (21.14), T_{44} involves μ and it seems reasonable, therefore, to expect that the other member of the equation expressing the new law of gravitation will provide terms which can receive an approximate interpretation as $\nabla^2 U$. This implies that second-order derivatives of the metric tensor components will probably be present. We therefore have a requirement for a second rank contravariant symmetric tensor involving second-order derivatives of the g_{ij} and of vanishing divergence to which T^{ij} can be assumed proportional. Einstein's tensor (42.20) possesses these characteristics and consequently we shall put

$$R^{ij} - \tfrac{1}{2}g^{ij}R = -\kappa T^{ij} \tag{47.11}$$

where κ is a constant of proportionality which must be related to G and which we shall later prove to be positive. Equation (47.11) expresses *Einstein's law of gravitation*; by lowering the indices successively, it may be expressed in the two alternative forms

$$R^i_j - \tfrac{1}{2}\delta^i_j R = -\kappa T^i_j \tag{47.12}$$

$$R_{ij} - \tfrac{1}{2}g_{ij}R = -\kappa T_{ij} \tag{47.13}$$

If equation (47.12) is contracted, it is found that

$$R = \kappa T \tag{47.14}$$

where $T = T^i_i$. It now follows that Einstein's law of gravitation can also be expressed in the form

$$R_{ij} = \kappa(\tfrac{1}{2}g_{ij}T - T_{ij}) \tag{47.15}$$

with two other forms obtained by raising subscripts.

Since the divergence of g^{ij} vanishes, a possible alternative to the law (47.11) is

$$R^{ij} - \tfrac{1}{2}g^{ij}R - \Lambda g^{ij} = -\kappa T^{ij} \tag{47.16}$$

where Λ is a constant. The law (47.11) gives results which agree with observation over regions of space of galactic dimensions, so that it is certain that, even if Λ is not zero, it is exceedingly small. However, the extra term has entered into some cosmological investigations (see Chapter 7).

48. Acceleration of a particle in a weak gravitational field

In a gravitational field, such as the one due to the earth, the geometry of space is not Euclidean and no truly inertial frame exists. In spite of this, we experience no practical difficulty in establishing rectangular Cartesian axes $Oxyz$ at the earth's surface relative to which for all practical purposes the geometry is Euclidean and the behaviour of electromagnetic systems is indistinguishable from their behaviour in an inertial frame. It must be concluded, therefore, that such a gravitational field is comparatively weak and hence that, with respect to such axes and their associated clocks, the space–time metric will not differ greatly from that given by equation (45.5). Putting

$$x^1 = x, \quad x^2 = y, \quad x^3 = z, \quad x^4 = ict, \tag{48.1}$$

in terms of the x^i the metric will be given by

$$ds^2 = dx^i dx^i \tag{48.2}$$

approximately. With respect to the x^i-frame, it will accordingly be assumed that

$$g_{ij} = \delta_{ij} + h_{ij} \tag{48.3}$$

where the δ_{ij} are Kronecker deltas and the h_{ij} are small by comparison.

Consider a particle moving in a weak gravitational field whose metric tensor is given by equation (48.3). The contravariant metric tensor will be given by an equation of the form

$$g^{ij} = \delta^{ij} + k^{ij} \tag{48.4}$$

where the k^{ij} are of the same order of smallness as the h_{ij}. Then, since

$$[ij,k] = \frac{1}{2}\left(\frac{\partial h_{jk}}{\partial x^i} + \frac{\partial h_{ik}}{\partial x^j} - \frac{\partial h_{ij}}{\partial x^k}\right) \tag{48.5}$$

it follows that, to a first approximation,

$$\{_i{}^k{}_j\} = \delta^{kr}[ij,r] = \frac{1}{2}\left(\frac{\partial h_{jk}}{\partial x^i} + \frac{\partial h_{ik}}{\partial x^j} - \frac{\partial h_{ij}}{\partial x^k}\right) \tag{48.6}$$

The equations of motion of the particle can now be written down as at (46.5). By equation (46.2),

$$\frac{dx^i}{ds} = \frac{dx^i}{dt}\frac{dt}{ds} = (v^2 - c^2)^{-1/2}\,(\mathbf{v}, ic) \tag{48.7}$$

where $\mathbf{v}$ is the particle's velocity in the quasi-inertial frame. Hence, if the particle is stationary in the frame at the instant under consideration,

$$\frac{dx^i}{ds} = (\mathbf{0}, 1) \tag{48.8}$$

and the equations of motion (46.5) reduce to the form

$$\frac{d^2x^i}{ds^2} + \{_4{}^i{}_4\} = 0 \tag{48.9}$$

correct to the first order in the h_{ij}. Substituting from equation (48.6), this is seen to be equivalent to

$$\frac{d^2x^i}{ds^2} = \frac{1}{2}\frac{\partial h_{44}}{\partial x^i} - \frac{\partial h_{4i}}{\partial x^4} \tag{48.10}$$

Differentiating equation (48.7) with respect to s and making use of equation (46.2), we obtain

$$\frac{d^2x^i}{ds^2} = (v^2 - c^2)^{-1}\left(\frac{d\mathbf{v}}{dt}, 0\right) - v\frac{dv}{dt}(v^2 - c^2)^{-2}\,(\mathbf{v}, ic) \tag{48.11}$$

and, when $\mathbf{v} = \mathbf{0}$, this reduces to

$$\frac{d^2x^i}{ds^2} = -\frac{1}{c^2}\left(\frac{d\mathbf{v}}{dt}, 0\right) \tag{48.12}$$

From equations (48.10) and (48.12), we deduce that the components of the acceleration of the stationary particle in the directions of the rectangular axes are

$$-c^2 \left(\frac{1}{2} \frac{\partial h_{44}}{\partial x^i} - \frac{\partial h_{4i}}{\partial x^4} \right) \tag{48.13}$$

for $i = 1, 2, 3$. Reverting to the original coordinates (x, y, z, t), these components are written

$$-c^2 \left(\frac{1}{2} \frac{\partial h_{44}}{\partial x} + \frac{i}{c} \frac{\partial h_{41}}{\partial t} \right), \qquad \text{etc} \tag{48.14}$$

Hence, if the field does not vary with the time, the acceleration vector is

$$- \text{grad} \left(\tfrac{1}{2} c^2 h_{44} \right) \tag{48.15}$$

But, if U is the Newtonian potential function for the field, this acceleration will be $- \text{grad} U$. It follows that, for a weak field, a Newtonian scalar potential U exists and is related to the space–time metric by the equation

$$U = \tfrac{1}{2} c^2 h_{44} \tag{48.16}$$

Alternatively, we can write

$$g_{44} = 1 + \frac{2U}{c^2} \tag{48.17}$$

49. Newton's law of gravitation

In this section it will be shown that Newton's law of gravitation may be deduced from Einstein's law in the normal case when the gravitational field's intensity is weak and the matter distribution is static.

First consider the form taken by the Riemann–Christoffel tensor in the space–time of a weak field. In the x^i-frame, the metric tensor is given by equation (48.3) and the Christoffel three-index symbols by equation (48.6). If products of the h_{ij} are to be neglected, equation (36.21) shows that

$$B^i_{jkl} = \frac{\partial}{\partial x^k} \{^{\,i}_{jl}\} - \frac{\partial}{\partial x^i} \{^{\,i}_{jk}\} \tag{49.1}$$

approximately. Hence the Ricci tensor is given by

$$
\begin{aligned}
R_{jk} &= \frac{\partial}{\partial x^k} \{^{\,i}_{ji}\} - \frac{\partial}{\partial x^i} \{^{\,i}_{jk}\} \\
&= \frac{1}{2} \frac{\partial}{\partial x^k} \left\{ \frac{\partial h_{ji}}{\partial x^i} + \frac{\partial h_{ii}}{\partial x^j} - \frac{\partial h_{ij}}{\partial x^i} \right\} \\
&\quad - \frac{1}{2} \frac{\partial}{\partial x^i} \left\{ \frac{\partial h_{ij}}{\partial x^k} + \frac{\partial h_{ki}}{\partial x^j} - \frac{\partial h_{jk}}{\partial x^i} \right\}
\end{aligned}
$$

$$= \frac{1}{2} \left\{ \frac{\partial^2 h_{ii}}{\partial x^j \partial x^k} + \frac{\partial^2 h_{jk}}{\partial x^i \partial x^i} - \frac{\partial^2 h_{ij}}{\partial x^i \partial x^k} - \frac{\partial^2 h_{ki}}{\partial x^i \partial x^j} \right\} \tag{49.2}$$

In particular, putting $j = k = 4$, we find that

$$R_{44} = \frac{1}{2} \left\{ \frac{\partial^2 h_{ii}}{\partial x^4 \partial x^4} + \frac{\partial^2 h_{44}}{\partial x^i \partial x^i} - 2 \frac{\partial^2 h_{i4}}{\partial x^i \partial x^4} \right\} \tag{49.3}$$

If the matter distribution is static in the quasi-inertial frame being employed, the h_{ij} will be independent of t and equation (49.3) reduces to

$$R_{44} = \tfrac{1}{2} \nabla^2 h_{44} \tag{49.4}$$

where $\nabla^2 = \partial^2/\partial x^2 + \partial^2/\partial y^2 + \partial^2/\partial z^2$. If U is the Newtonian potential for the field, equation (48.16) now shows that

$$R_{44} = \frac{1}{c^2} \nabla^2 U \tag{49.5}$$

Assuming that no electromagnetic field is present and that the contribution to the energy–momentum tensor of any stress forces within the matter distribution responsible for the gravitational field is negligible, T_{ij} will be determined by equation (21.16). But, since the distribution is static, its 4-velocity of flow $\mathbf{V}$ at every point is $(\mathbf{0}, ic)$ and hence all components of T_{ij}, with the exception of T_{44}, are zero. In this case,

$$T_{44} = -c^2 \mu_{00} \tag{49.6}$$

where μ_{00}, for zero velocity of matter, is the ordinary mass density. Also

$$T = T_i^i = T_{ii} = T_{44} = -c^2 \mu_{00} \tag{49.7}$$

The 44-component of Einstein's gravitation law in the form of equation (47.15) can now be expressed approximately

$$\frac{1}{c^2} \nabla^2 U = \tfrac{1}{2} \kappa c^2 \mu_{00}$$

or

$$\nabla^2 U = \tfrac{1}{2} \kappa c^4 \mu_{00} \tag{49.8}$$

This is the Poisson equation (47.10) of classical Newtonian theory, provided we accept

$$\kappa = \frac{8\pi G}{c^4} \tag{49.9}$$

This specifies κ in terms of the gravitational constant.

50. Freely falling dust cloud

Consider the case of a cloud of particles falling freely in the field of the cloud itself, there being no other forces present in the system. The energy–momentum tensor for such an incoherent cloud has been calculated relative to an inertial frame as equation (21.16). This equation will be taken to provide a definition of T_{ij} in the freely falling frame at any point of the cloud.

In an arbitrary x-frame, let $x^i = x^i(\tau)$ be parametric equations of the world-line of some particle of the cloud, τ being the proper time measured by a standard clock moving with the particle. Then the 4-velocity of flow of the particle at time τ is defined by the equation

$$V^i = \frac{dx^i}{d\tau} = ic\frac{dx^i}{ds} \tag{50.1}$$

where s is the interval parameter measured along the world-line. The square of the magnitude of the 4-velocity of flow is

$$V^2 = g_{ij}V^iV^j = -c^2 g_{ij}\frac{dx^i}{ds}\frac{dx^j}{ds} = -c^2 \tag{50.2}$$

This equation can also be written

$$V_iV^i = -c^2 \tag{50.3}$$

Now consider the tensor equation

$$T^{ij} = \mu_{00}V^iV^j \tag{50.4}$$

where μ_{00} is the mass density as measured in a freely falling frame moving with the cloud; μ_{00} is clearly a 4-invariant. In any freely falling frame, this equation reduces to equation (21.16) and is accordingly valid; this establishes its validity in all frames.

Equation (47.4) is known to follow from Einstein's equation of gravitation and, in this case, takes the form

$$(\mu_{00}V^iV^j)_{;j} = (\mu_{00}V^j)_{;j}V^i + \mu_{00}V^jV^i_{;j} = 0 \tag{50.5}$$

Multiplication by V_i now gives

$$(\mu_{00}V^j)_{;j}V^iV_i + \mu_{00}V^i_{;j}V_iV^j = 0 \tag{50.6}$$

Differentiating equation (50.3) with respect to x^j, we find

$$V_{i;j}V^i + V_iV^i_{;j} = 0 \tag{50.7}$$

Raising and lowering the index i in the two factors of the first term, this equation is seen to be equivalent to

$$V^i_{;j}V_i = 0 \tag{50.8}$$

Equation (50.6) accordingly reduces to

$$(\mu_{00}V^j)_{;j} = 0 \tag{50.9}$$

Equation (50.5) now gives

$$V^i_{;j}V^j = 0 \tag{50.10}$$

or

$$\left(\frac{\partial V^i}{\partial x^j} + \Gamma^i_{kj} V^k\right) V^j = 0 \tag{50.11}$$

Hence

$$\frac{\partial V^i}{\partial x^j}\frac{\mathrm{d}x^j}{\mathrm{d}\tau} + \Gamma^i_{kj}\frac{\mathrm{d}x^k}{\mathrm{d}\tau}\frac{\mathrm{d}x^j}{\mathrm{d}\tau} = 0 \tag{50.12}$$

or

$$\frac{\mathrm{d}^2 x^i}{\mathrm{d}\tau^2} + \Gamma^i_{kj}\frac{\mathrm{d}x^k}{\mathrm{d}\tau}\frac{\mathrm{d}x^j}{\mathrm{d}\tau} = 0 \tag{50.13}$$

This can also be written

$$\frac{\mathrm{d}^2 x^i}{\mathrm{d}s^2} + \Gamma^i_{kj}\frac{\mathrm{d}x^k}{\mathrm{d}s}\frac{\mathrm{d}x^j}{\mathrm{d}s} = 0 \tag{50.14}$$

a result which proves that the world-lines of the particles of the cloud are geodesics.

That the world-lines of freely falling particles in general circumstances are geodesics can be derived from equation (47.4), proving that Einstein's law includes its own law of motion for a particle in a gravitational field.

51. Metrics with spherical symmetry

When a change is made in the space–time coordinate frame from coordinates x^i to coordinates $\bar{x}^i$, the metric tensor g_{ij} will change to $\bar{g}_{ij}$ by the law of transformation of a covariant tensor. In general, the g_{ij} will be functions of the x^i and the $\bar{g}_{ij}$ will be functions of the $\bar{x}^i$, but it will not usually be the case that the $\bar{g}_{ij}$ are the *same* functions of the 'barred' coordinates that the g_{ij} are of the 'unbarred' coordinates, i.e., the functions $g_{ij}(x^k)$ are not *form invariant* under general coordinate transformations. However, in some special cases, it is possible for these functions to be form invariant under a whole group of transformations, and we shall study such a case in this section.

In a gravitational field, the geometry can only be quasi-Euclidean and consequently rectangular Cartesian axes do not exist. Nevertheless, no difficulty is experienced in practice in defining such axes approximately and we shall suppose, therefore, that the coordinates x, y, z, t of an event in the gravitational field about to be considered are interpreted physically as rectangular Cartesian coordinates and time. We shall now search for a metric which, when expressed in those coordinates, is form invariant with respect to the group of coordinate transformations which will be interpreted physically as rotations of the rectangular axes $Oxyz$ (t is to remain unaltered). To be precise, it will be supposed that spatial coordinates (x, y, z) can be defined such that the metric $g_{ij}(x, y, z, t)$ is form invariant under the group of orthogonal transformations $\bar{\mathbf{x}} = \mathbf{A}\mathbf{x}$, where $\mathbf{x} = (x, y, z)^T$, $\bar{\mathbf{x}} = (\bar{x}, \bar{y}, \bar{z})^T$ and $\mathbf{A}\mathbf{A}^T = \mathbf{I}$. Such a metric will be said to be *spherically symmetric* about O.

Invariants for this group of coordinate transformations, which are of degree no higher than the second in the coordinate differentials dx, dy, dz, are

$$x^2 + y^2 + z^2, \quad x\,dx + y\,dy + z\,dz, \quad dx^2 + dy^2 + dz^2 \tag{51.1}$$

Introducing spherical polar coordinates (r, θ, ϕ), which will be *defined* by the equations (30.3), these invariants may be written

$$r^2, \quad r\,dr, \quad dr^2 + r^2\,d\theta^2 + r^2\sin^2\theta\,d\phi^2. \tag{51.2}$$

It follows that
$$r, \quad dr, \quad d\theta^2 + \sin^2\theta\,d\phi^2 \tag{51.3}$$

are invariants. The most general metric with spherical symmetry can now be built up in the form

$$ds^2 = A(r,t)\,dr^2 + B(r,t)\,(d\theta^2 + \sin^2\theta\,d\phi^2)$$
$$+ C(r,t)\,dr\,dt + D(r,t)\,dt^2 \tag{51.4}$$

We now replace r by a new coordinate r' according to the transformation equation
$$r'^2 = B(r,t) \tag{51.5}$$

Then
$$ds^2 = E(r',t)\,dr'^2 + r'^2\,(d\theta^2 + \sin^2\theta\,d\phi^2)$$
$$+ F(r',t)\,dr'\,dt + G(r',t)\,dt^2 \tag{51.6}$$

In a truly inertial frame, spherical polar coordinates can be defined exactly and the metric will, by equation (45.5), be expressed in the form

$$ds^2 = dr^2 + r^2(d\theta^2 + \sin^2\theta\,d\phi^2) - c^2\,dt^2 \tag{51.7}$$

Comparing equations (51.6) and (51.7), it is clear that in a region for which (51.6) is the metric, r' will behave approximately like a true spherical polar coordinate r. We shall accordingly drop the primes and write

$$ds^2 = E(r,t)\,dr^2 + r^2(d\theta^2 + \sin^2\theta\,d\phi^2)$$
$$+ F(r,t)\,dr\,dt + G(r,t)\,dt^2 \tag{51.8}$$

If our frame is quasi-inertial, equation (51.7) must be an approximation for equation (51.8) and the following equations must therefore be true approximately:

$$E(r,t) = 1, \quad F(r,t) = 0, \quad G(r,t) = -c^2 \tag{51.9}$$

Consider now the special case when the gravitational field is static in the quasi-inertial frame for which (r, θ, ϕ) are approximate spherical polar coordinates and t is the time. The functions E, F, G will then be independent of t. Also, space–time will be symmetric as regards past and future senses of the time variable and this implies that ds^2 is unaltered when dt is replaced by $-dt$. Thus $F = 0$ and we have

$$ds^2 = a\,dr^2 + r^2(d\theta^2 + \sin^2\theta\,d\phi^2) - bc^2\,dt^2 \tag{51.10}$$

where a, b are functions of r both approximating unity in a weak field.

At any fixed instant t, the metric of space in the presence of this gravitational field can be obtained from the last equation by putting $dt = 0$. Thus, it is

$$ds^2 = a\,dr^2 + r^2(d\theta^2 + \sin^2\theta\,d\phi^2) \tag{51.11}$$

Consider the 'circle' $r = r_0$ in the 'plane' $\theta = \frac{1}{2}\pi$. The length of the element of the circle with end points (r_0, ϕ), $(r_0, \phi + d\phi)$, has length $ds = r_0\,d\phi$. Thus the total length of the circle is $2\pi r_0$. However, r_0 will not be the length of a radius $\phi = \phi_0$ of this circle for, if an element of such a radius has end points (r, ϕ_0), $(r + dr, \phi_0)$, the length of the element is $ds = a^{1/2}\,dr$ and the total length of the radius is accordingly

$$\int_0^{r_0} a^{1/2}\,dr \tag{51.12}$$

Clearly, the Euclidean formula for the circumference of a circle of given radius does not apply.

For the metric (51.10), taking

$$x^1 = r, \quad x^2 = \theta, \quad x^3 = \phi, \quad x^4 = t \tag{51.13}$$

we have

$$g_{11} = a, \quad g_{22} = r^2, \quad g_{33} = r^2\sin^2\theta, \quad g_{44} = -bc^2 \tag{51.14}$$

all other g_{ij} being zero. Thus

$$g = -abc^2 r^4 \sin^2\theta \tag{51.15}$$

and hence

$$g^{11} = \frac{1}{a}, \quad g^{22} = \frac{1}{r^2}, \quad g^{33} = \frac{1}{r^2\sin^2\theta}, \quad g^{44} = -\frac{1}{bc^2} \tag{51.16}$$

all other g^{ij} being zero. The three-index symbols can now be calculated and, putting $a = e^\alpha$, $b = e^\beta$, those which do not vanish are listed below:

$$
\begin{aligned}
&\left\{ \begin{matrix} 1 \\ 1 \quad 1 \end{matrix} \right\} = \tfrac{1}{2}\alpha' \\[6pt]
&\left\{ \begin{matrix} 2 \\ 1 \quad 2 \end{matrix} \right\} = \left\{ \begin{matrix} 2 \\ 2 \quad 1 \end{matrix} \right\} = 1/r \\[6pt]
&\left\{ \begin{matrix} 3 \\ 1 \quad 3 \end{matrix} \right\} = \left\{ \begin{matrix} 3 \\ 3 \quad 1 \end{matrix} \right\} = 1/r \\[6pt]
&\left\{ \begin{matrix} 4 \\ 1 \quad 4 \end{matrix} \right\} = \left\{ \begin{matrix} 4 \\ 4 \quad 1 \end{matrix} \right\} = \tfrac{1}{2}\beta' \\[6pt]
&\left\{ \begin{matrix} 1 \\ 2 \quad 2 \end{matrix} \right\} = -re^{-\alpha}
\end{aligned}
\tag{51.17}
$$

$$\begin{Bmatrix} 3 \\ 2\ 3 \end{Bmatrix} = \begin{Bmatrix} 3 \\ 3\ 2 \end{Bmatrix} = \cot\theta$$

$$\begin{Bmatrix} 1 \\ 3\ 3 \end{Bmatrix} = -re^{-\alpha}\sin^2\theta$$

$$\begin{Bmatrix} 2 \\ 3\ 3 \end{Bmatrix} = -\sin\theta\cos\theta$$

$$\begin{Bmatrix} 1 \\ 4\ 4 \end{Bmatrix} = \tfrac{1}{2}c^2\beta'e^{\beta-\alpha}$$

primes denoting differentiations with respect to r.

The non-zero components of the Ricci tensor are now calculated to be:

$$R_{11} = \tfrac{1}{2}\beta'' + \tfrac{1}{4}\beta'^2 - \tfrac{1}{4}\alpha'\beta' - \frac{1}{r}\alpha'$$

$$R_{22} = e^{-\alpha}(\tfrac{1}{2}r\beta' - \tfrac{1}{2}r\alpha' + 1) - 1$$

$$R_{33} = R_{22}\sin^2\theta$$

$$R_{44} = c^2 e^{\beta-\alpha}(-\tfrac{1}{2}\beta'' - \tfrac{1}{4}\beta'^2 + \tfrac{1}{4}\alpha'\beta' - \frac{1}{r}\beta')$$

$$(51.18)$$

52. Schwarzschild's solution

The static, spherically symmetrical metric (51.10) will determine the gravitational field of a static distribution of matter also having spherical symmetry, provided it satisfies Einstein's equations (47.15). We shall consider the special case when the whole of space is devoid of matter, apart from a spherical body with its centre at the centre of symmetry O. Then $T_{ij} = 0, T = 0$ at all points outside the body and Einstein's equations reduce in this region to

$$R_{ij} = 0 \qquad (52.1)$$

By equations (51.18), these are satisfied by the metric (51.10), provided

$$\beta'' + \tfrac{1}{2}\beta'^2 - \tfrac{1}{2}\alpha'\beta' - \frac{2}{r}\alpha' = 0 \qquad (52.2)$$

$$\tfrac{1}{2}r\beta' - \tfrac{1}{2}r\alpha' + 1 = e^\alpha \qquad (52.3)$$

$$\beta'' + \tfrac{1}{2}\beta'^2 - \tfrac{1}{2}\alpha'\beta' + \frac{2}{r}\beta' = 0 \qquad (52.4)$$

Subtracting equation (52.2) from (52.4), it follows that

$$\alpha + \beta = \text{constant} \qquad (52.5)$$

But, as $r \to \infty$, we shall assume that our metric approaches that given by equation (51.7), valid in the absence of a gravitational field. Thus, at infinity, $\alpha = \beta = 0$ and hence

$$\alpha + \beta = 0 \tag{52.6}$$

Eliminating β from equation (52.3), it will be found that

$$r\alpha' = 1 - e^{\alpha} \tag{52.7}$$

The variables are separable and this equation is easily integrated to yield

$$a = e^{\alpha} = (1 - 2m/r)^{-1} \tag{52.8}$$

where m is a constant of integration. Then

$$b = e^{\beta} = 1 - 2m/r \tag{52.9}$$

and it may be verified that each of the equations (52.2)–(52.4) is satisfied by these expressions for α and β.

We have accordingly arrived at a metric

$$ds^2 = \frac{dr^2}{1 - (2m/r)} + r^2(d\theta^2 + \sin^2\theta \, d\phi^2) - c^2\left(1 - \frac{2m}{r}\right)dt^2 \tag{52.10}$$

which is spherically symmetrical and can represent the gravitational field outside a spherical body with its centre at the pole of spherical polar coordinates (r, θ, ϕ). This was first obtained by Schwarzschild. It will be proved in the next section that the constant m is proportional to the mass of the body. This may also be deduced from equation (48.17), for the potential U at a distance r from a spherical body of mass M is given by

$$U = -\frac{GM}{r} \tag{52.11}$$

and hence

$$g_{44} = 1 - \frac{2GM}{c^2 r} \tag{52.12}$$

Now g_{44} is the coefficient of $(dx^4)^2 = -c^2 \, dt^2$ in the metric and hence

$$b = 1 - \frac{2GM}{c^2 r}. \tag{52.13}$$

Comparing equations (52.9) and (52.13), it will be seen that

$$m = \frac{GM}{c^2} \tag{52.14}$$

It is clear from equation (52.10) that the metric is not valid for $r = 2m = 2GM/c^2$. This is the *Schwarzschild radius*. In SI units, $c = 3 \times 10^8$ and for the earth $GM = 3.991 \times 10^{14}$, so that the radius for this body is about 9 mm; since the metric is only applicable in the region outside the earth, no difficulty is encountered in this case. However, for an exceptionally dense body, the radius may extend into the surrounding space and, for values of r less than the radius, the metric needs special consideration (see section 57).

53. Planetary orbits

The attractions of the planets upon the sun cause this body to have a small acceleration relative to an inertial frame. If, therefore, a coordinate frame moving with the sun is constructed, relative to this frame there will be a gravitational field corresponding to this acceleration in addition to that of the sun and planets. However, for the purpose of the following analysis, this field and the fields of the planets will be neglected. Thus, relative to spherical polar coordinates having their pole at the centre of the sun, the gravitational field will be assumed determined by the Schwarzschild metric (52.10). The planets will be treated as particles possessing negligible gravitational fields, whose world-lines are geodesics in space–time. We proceed to calculate these geodesics.

Since the intervals between adjacent points on the world-line of a particle are necessarily timelike, s will be purely imaginary along such a curve. When calculating geodesics it is usually more convenient, therefore, to replace s by τ and to work from the metric expression for $d\tau^2$. Thus, in this section, the Schwarzschild metric will be taken in the form

$$d\tau^2 = -\frac{1}{c^2}\left(\frac{dr^2}{1-2m/r} + r^2(d\theta^2 + \sin^2\theta\, d\phi^2)\right) + (1 - 2m/r)\, dt^2 \qquad (53.1)$$

and equations (43.19) for the geodesics become

$$\frac{d}{d\tau}\left(\frac{r}{r-2m}\frac{dr}{d\tau}\right) + \frac{m}{(r-2m)^2}\left(\frac{dr}{d\tau}\right)^2 - r\left(\frac{d\theta}{d\tau}\right)^2 - r\sin^2\theta\left(\frac{d\phi}{d\tau}\right)^2 + \frac{mc^2}{r^2}\left(\frac{dt}{d\tau}\right)^2 = 0$$

$$(53.2)$$

$$\frac{d}{d\tau}\left(r^2\frac{d\theta}{d\tau}\right) - r^2\sin\theta\cos\theta\left(\frac{d\phi}{d\tau}\right)^2 = 0 \qquad (53.3)$$

$$\frac{d}{d\tau}\left(r^2\sin^2\theta\frac{d\phi}{d\tau}\right) = 0 \qquad (53.4)$$

$$\frac{d}{d\tau}\left(\frac{r-2m}{r}\frac{dt}{d\tau}\right) = 0 \qquad (53.5)$$

The first of this set of equations will be replaced by the first integral (43.6), viz.

$$\frac{r}{r-2m}\left(\frac{dr}{d\tau}\right)^2 + r^2\left\{\left(\frac{d\theta}{d\tau}\right)^2 + \sin^2\theta\left(\frac{d\phi}{d\tau}\right)^2\right\} - \frac{c^2}{r}(r-2m)\left(\frac{dt}{d\tau}\right)^2 = -c^2$$

$$(53.6)$$

We now choose the spherical polar coordinates so that the planet is moving initially in the plane $\theta = \frac{1}{2}\pi$. Then $d\theta/d\tau = 0$ initially and hence, by equation (53.3), $d^2\theta/d\tau^2 = 0$ at this instant. By repeated differentiation of this equation and substitution of initial values, it is found that all derivatives of θ vanish initially. Hence, by Maclaurin's theorem, $\theta = \frac{1}{2}\pi$ for all values of τ, proving that the planet continues to move in the 'plane' $\theta = \frac{1}{2}\pi$ indefinitely.

Integrating equations (53.4) and (53.5), and putting $\theta = \frac{1}{2}\pi$, we get

$$\frac{d\phi}{d\tau} = \frac{h}{r^2} \tag{53.7}$$

$$\frac{dt}{d\tau} = \frac{kr}{r-2m} \tag{53.8}$$

where h and k are constants of integration.

Substituting for $d\phi/d\tau$, $dt/d\tau$ from the last two equations and putting $\theta = \frac{1}{2}\pi$ in equation (53.6), it follows that

$$\left(\frac{dr}{d\tau}\right)^2 + \frac{h^2}{r^3}(r-2m) = c^2(k^2-1) + \frac{2mc^2}{r} \tag{53.9}$$

Then, eliminating $d\tau$ between this equation and equation (53.7), we obtain the equation for the orbit, viz.

$$\left(\frac{h}{r^2}\frac{dr}{d\phi}\right)^2 + \frac{h^2}{r^2} = c^2(k^2-1) + \frac{2mc^2}{r} + \frac{2mh^2}{r^3} \tag{53.10}$$

With $u = 1/r$, this reduces to the form

$$\left(\frac{du}{d\phi}\right)^2 + u^2 = \frac{c^2}{h^2}(k^2-1) + \frac{2mc^2}{h^2}u + 2mu^3 \tag{53.11}$$

Differentiating through with respect to ϕ, this equation takes a form which is familiar in the theory of orbits, viz.

$$\frac{d^2u}{d\phi^2} + u = \frac{mc^2}{h^2} + 3mu^2 \tag{53.12}$$

The corresponding equation governing the orbit according to classical mechanics is

$$\frac{d^2u}{d\phi^2} + u = \frac{GM}{h^2} \tag{53.13}$$

where M is the mass of the attracting body and h is the constant velocity moment of the planet about the centre of attraction, i.e.

$$r^2\frac{d\phi}{dt} = h \tag{53.14}$$

If we identify the time variable t of classical theory with the proper time τ in the relativistic theory, equations (53.7) and (53.14) become identical and our choice of h for the constant in equation (53.7) is justified. Also, provided we take

$$m = \frac{GM}{c^2} \tag{53.15}$$

(confirming equation (52.14)), equation (53.12) corresponds to the classical equation (53.13), although there is now an additional term $3mu^2$. The ratio of the additional term $3mu^2$ to the 'inverse square law' term mc^2/h^2 is

$$\frac{3h^2u^2}{c^2} = \frac{3}{c^2}r^2\dot{\phi}^2 \tag{53.16}$$

by equation (53.14). $r\dot{\phi}$ is the transverse component of the planet's velocity and, for the planets of the solar system, takes its largest value in the case of Mercury, viz. 4.8×10^4 m/s. Since $c = 3 \times 10^8$ m/s, the ratio of the terms is in this case $7 \cdot 7 \times 10^{-8}$, which is very small. However, the effect of the additional term proves to be cumulative, as will now be proved, and for this reason an observational check can be made.

The solution of the classical equation (53.13), viz.

$$u = \frac{\mu}{h^2}\{1 + e\cos(\phi - \tilde{\omega})\} \tag{53.17}$$

where $\mu = GM = mc^2$, e is the eccentricity of the orbit and $\tilde{\omega}$ is the longitude of perihelion, will be an approximate, though highly accurate, solution of equation (53.12). Hence the error involved in taking

$$3mu^2 = \frac{3m\mu^2}{h^4}\{1 + e\cos(\phi - \tilde{\omega})\}^2 \tag{53.18}$$

will be absolutely inappreciable, since this term is very small in any case. Equation (53.12) can accordingly be replaced by

$$\frac{d^2u}{d\phi^2} + u = \frac{\mu}{h^2} + \frac{3m\mu^2}{h^4}\{1 + e\cos(\phi - \tilde{\omega})\}^2 \tag{53.19}$$

This equation will possess a solution of the form (53.17) with additional 'particular integral' terms corresponding to the new term (53.18). These prove to be as follows:

$$\frac{3m\mu^2}{h^4}\{1 + \tfrac{1}{2}e^2 - \tfrac{1}{6}e^2\cos 2(\phi - \tilde{\omega}) + e\phi\sin(\phi - \tilde{\omega})\} \tag{53.20}$$

The constant term cannot be observationally separated from that already occurring in equation (53.17). The term in $\cos 2(\phi - \tilde{\omega})$ has amplitude too small for detection. However, the remaining term has an amplitude which increases with ϕ and its effect is accordingly cumulative. Adding this to the solution (53.17), we obtain

$$u = \frac{\mu}{h^2} \left\{ 1 + e\cos(\phi - \tilde{\omega}) + \frac{3m\mu e}{h^2} \phi \sin(\phi - \tilde{\omega}) \right\}$$

$$= \frac{\mu}{h^2} \{ 1 + e\cos(\phi - \tilde{\omega} - \delta\tilde{\omega}) \} \tag{53.21}$$

where $\delta\tilde{\omega} = 3m\mu\phi/h^2$ and we have neglected terms $O(\delta\tilde{\omega}^2)$.

Equation (53.21) indicates that the longitude of perihelion should steadily increase according to the equation

$$\delta\tilde{\omega} = \frac{3m\mu}{h^2} \phi = \frac{3\mu^2}{c^2 h^2} \phi = \frac{3\mu}{c^2 l} \phi \tag{53.22}$$

where $l = h^2/\mu$ is the semi-latus rectum of the orbit. Taking $\mu = 1 \cdot 33 \times 10^{20}$ SI units for the sun, $c = 3 \times 10^8$ and $l = 5 \cdot 79 \times 10^{10}$ for Mercury, it will be found that the predicted angular advance of perihelion per century for this planet's orbit is 43″. This is in agreement with the observed value. The advances predicted for the other planets are too small to be observable at the present time.

54. Gravitational deflection of a light ray

In section 7 it was shown that the proper time interval between the transmission of a light signal and its reception at a distant point is zero. It was there assumed that the signal was being propagated in an inertial frame and hence that no gravitational field was present. This result can be expressed by saying that

$$ds = 0 \tag{54.1}$$

for any two neighbouring points on the world-line of a light signal. Now, null geodesics in the space–time having metric (45.5) are defined by equation (54.1) and the equations

$$\frac{d^2 x}{d\lambda^2} = \frac{d^2 y}{d\lambda^2} = \frac{d^2 z}{d\lambda^2} = \frac{d^2 t}{d\lambda^2} \tag{54.2}$$

for the three index symbols are all zero. Equations (54.2) imply that along a null geodesic x, y, z are linearly dependent upon t. But this is certainly true for the coordinates of a light signal being propagated in an inertial frame. We conclude that the world-lines of light signals are null geodesics in space–time, in this case.

Since an inertial frame can always be found for a sufficiently small space–time region even in the presence of a gravitational field, it follows that the world-line of a light signal in any such region is a null geodesic. We shall accept the obvious generalization of this result, viz. that the world-lines of light signals over an unlimited region of space–time are null geodesics.

We shall now employ this principle to calculate the path of a light ray in the gravitational field of a spherical body. Taking the space–time metric in the Schwarzschild form (equation (52.10)), the equations governing a null geodesic are identical with the equations (53.2)–(53.5) after τ has been replaced by λ. The first integral (43.13) takes the form

$$\frac{r}{r-2m}\left(\frac{dr}{d\lambda}\right)^2 + r^2\left\{\left(\frac{d\theta}{d\lambda}\right)^2 + \sin^2\theta\left(\frac{d\phi}{d\lambda}\right)^2\right\} - \frac{c^2}{r}(r-2m)\left(\frac{dt}{d\lambda}\right)^2 = 0 \tag{54.3}$$

Without loss of generality, we shall again put $\theta = \frac{1}{2}\pi$, so that a ray in the equatorial plane is being considered and then proceed exactly as in the last section to derive the equation

$$\frac{d^2u}{d\phi^2} + u = 3mu^2 \tag{54.4}$$

where $u = 1/r$. This equation determines the family of light rays in the equatorial plane.

As a first approximation to the solution of equation (54.4), we shall neglect the right-hand member. Then

$$u = \frac{1}{R}\cos(\phi + \alpha) \tag{54.5}$$

where R, α are constants of integration. This is the polar equation of a straight line whose perpendicular distance from the centre of attraction is R. As might have been expected, therefore, provided the gravitational field is not too intense, the light rays will be straight lines. This deduction is, of course, confirmed by observation. Thus, as the moon's motion causes its disc to approach the position of a star on the celestial sphere and ultimately to occult this body, no appreciable deflection of the position of the star on the celestial sphere can be detected.

Again, without loss of generality, we shall put $\alpha = 0$ so that the light ray, as given by equation (54.5), is parallel to the y-axis ($\phi = \pm\frac{1}{2}\pi$). Then, putting $u = \cos\phi/R$ in the right-hand member of equation (54.4), this becomes

$$\frac{d^2u}{d\phi^2} + u = \frac{3m}{R^2}\cos^2\phi \tag{54.6}$$

The additional 'particular integral' term is now found to be

$$\frac{m}{R^2}(2 - \cos^2\phi) \tag{54.7}$$

and hence the second approximation to the polar equation of the light ray is

$$u = \frac{1}{R}\cos\phi + \frac{m}{R^2}(2 - \cos^2\phi) \tag{54.8}$$

At each end of the ray $u = 0$ and hence

$$\frac{m}{R}\cos^2\phi - \cos\phi - \frac{2m}{R} = 0 \tag{54.9}$$

Assuming m/R to be small, this quadratic equation has a small root and a large root. The small root is approximately

$$\cos \phi = -\frac{2m}{R} \tag{54.10}$$

and hence
$$\phi = \pm \left(\frac{\pi}{2} + \frac{2m}{R} \right) \tag{54.11}$$

at the two ends of the ray. The angular deflection in the ray caused by its passage through the gravitational field is accordingly

$$\frac{4m}{R} \tag{54.12}$$

approximately.

For a light ray grazing the sun's surface,

$$R = \text{sun's radius} = 6 \cdot 95 \times 10^8 \text{ m and } m = 1 \cdot 5 \times 10^3 \text{ m}$$

Thus the predicted deflection is $8 \cdot 62 \times 10^{-6}$ radians, or about $1 \cdot 77''$. This prediction has been checked by observing a star close to the sun's disc during a total eclipse. The experimental findings are in accord with the theoretical result.

55. Gravitational displacement of spectral lines

A *standard clock* will be taken to be any device which experiences a periodic motion, each cycle of which is indistinguishable from every other cycle. The passage of time between two events which occur in the neighbourhood of the clock is then measured by the number of cycles and fraction of a cycle which the device completes between these two instants. The clocks employed to determine the time coordinate ξ^4 of an event in section 45 were not, necessarily, standard clocks. Such coordinate clocks can have arbitrary variable rates, the only requirement being that, if A, B are two events in the vicinity of a coordinate clock and B occurs after A, then the coordinate-time for B must be greater than the coordinate-time for A.

The successive oscillations of atoms governing the motion of a modern atomic clock are indistinguishable from one another and it has been assumed that such a clock is being used whenever standard time is measured. The constancy of the rate of this fundamental physical process is not susceptible to experimental check, since it is the standard against which all other rates (e.g. the rate of rotation of the earth) are measured. By international agreement, one second is the time which elapses when a specific type of atomic system performs a specific number of oscillations and this definition applies in all regions of the cosmos and at all epochs; this fact should be borne in mind when phrases such as 'the first second after the big bang' occur in cosmological studies.

As explained in section 45, if x^i, $x^i + dx^i$ are the space–time coordinates of two adjacent events, then $d\tau = ds/ic$ is the time separating the events as measured by a standard clock which is present at both events. It is assumed that the interval between the events is timelike (i.e. $d\tau$ real) and that the clock is in a state of free fall during its passage from one event to the other; alternatively, if the clock is not freely falling, it is assumed that any effect on its rate of the gravitational field it experiences is corrected for.

Let x^i ($i = 1, 2, 3, 4$) be the coordinates of an event with respect to some space–time reference frame, x^1, x^2, x^3 being interpreted physically as spatial coordinates relative to a static frame and x^4/ic as time. If a standard clock is at rest relative to this frame, for adjacent points on its world-line $dx^1 = dx^2 = dx^3 = 0$ and hence

$$d\tau^2 = -ds^2/c^2 = -g_{44}(dx^4)^2/c^2 = g_{44}\,dt^2 \tag{55.1}$$

where we have put $x^4 = ict$. The time τ measured by the standard clock is therefore related to the time t shown on the coordinate clock at (x^1, x^2, x^3) by the equation

$$\tau = \int \sqrt{(g_{44})}\,dt \tag{55.2}$$

In the special case of the coordinate frame employed in section 48 which was stationary in a relatively weak static gravitational field, it was proved that g_{44} is given in terms of the Newtonian scalar potential U for the field by the approximate equation (48.17). Thus

$$d\tau = \left(1 + \frac{2U}{c^2}\right)^{1/2} dt \tag{55.3}$$

relates time intervals measured by a stationary standard clock and a coordinate clock at a point in a gravitational field where the potential is U. Now, when it is emitting its characteristic spectrum, an atom is operating as a standard clock. Consider, therefore, an atom for which the period (from standard tables) of one complete cycle of radiation corresponding to a certain spectral line is τ. If such an atom is stationary in the frame at a point P where the potential is U_1, the time for one complete cycle of the radiation as measured by a standard clock at the point will be τ and the coordinate-time for the cycle will be t, where

$$\tau = \sqrt{(1 + 2U_1/c^2)}\,t \tag{55.4}$$

Suppose this radiation is received at another fixed point Q where the potential is U_2. Let T be the difference between the coordinate-time of emission of light from P and the coordinate-time of its reception at Q; since the gravitational field is being assumed static and P, Q are fixed, T will be a fixed constant. Thus, if successive crests of the light wave are emitted from P at coordinate-times t_0, $t_0 + t$, these crests will be received at Q at coordinate-times $t_0 + T$, $t_0 + T + t$. It follows that the period of the radiation as measured by the coordinate clock at Q will also be t. However, a standard clock at Q will measure the period to be τ', where

$$\tau' = \sqrt{(1 + 2U_2/c^2)}\,t \tag{55.5}$$

Hence, if v is the standard frequency of the spectral line being observed and v' is the observed frequency of the line at Q, then

$$\frac{v'}{v} = \frac{\tau}{\tau'} = \left(\frac{1 + 2U_1/c^2}{1 + 2U_2/c^2}\right)^{1/2} \tag{55.6}$$

In particular, if $U_1 < U_2$, the observed light will have its frequency shifted towards the red.

In the case of an atom on the surface of the sun observed from a point on the earth's surface, it will be found that, in SI units,

$$U_1 = -1.914 \times 10^{11}, \quad U_2 = -9.512 \times 10^8$$

and thus

$$v' = 0.9999979\,v \tag{55.7}$$

This effect is so small that it is very difficult to measure. However, in the case of the companion of Sirius, the predicted effect is 30 times larger and has been confirmed by observation.

56. Maxwell's equations in a gravitational field

Over any sufficiently small region of space and restricted interval of time it is possible to define a rectangular Cartesian inertial frame, i.e. the frame in 'free fall' in the gravitational field. If the electric and magnetic components of the electromagnetic field are measured in this frame, the field tensor F_{ij} defined by equation (26.5) can be found. Employing the appropriate transformation equations, the components of this tensor relative to general coordinates x^i in the gravitational field can be computed. No distinction is made between covariant and contravariant properties relative to the original inertial frame so that, when transforming, F_{ij} may be treated as a covariant, contravariant or mixed tensor. If it is treated as a covariant tensor, the covariant components F_{ij} in the general x^i-frame will be generated. If it is treated as a contravariant or as a mixed tensor, the contravariant or mixed components F^{ij}, F^i_j respectively will be generated. In this way, the field tensor is defined at every point of space–time. Similarly, a current-density vector with covariant components J_i and contravariant components J^i is defined relative to the x^i-frame.

Consider the equations

$$F^{ij}_{\;\;;j} = \mu_0 J^i \tag{56.1}$$

$$F_{ij;k} + F_{jk;i} + F_{ki;j} = 0 \tag{56.2}$$

These are tensor equations and hence are valid in every space–time frame if they are valid in any one. But, relative to the inertial coordinate frame (x, y, z, ict)

which can be found for any sufficiently small space–time region, these equations reduce to equations (26.11) and hence are valid over such a region. Regarding the whole of space–time as an aggregate of such small elements, it follows that equations (56.1), (56.2) are universally true.

Since F^{ij} is skew-symmetric,

$$F^{ij}{}_{;j} = \frac{\partial F^{ij}}{\partial x^j} + \{^i_{rj}\} F^{rj} + \{^j_{rj}\} F^{ir}$$

$$= \frac{\partial F^{ij}}{\partial x^j} + \frac{1}{\sqrt{(-g)}} \frac{\partial}{\partial x^r} \{\sqrt{(-g)}\} F^{ir}$$

$$= \frac{1}{\sqrt{(-g)}} \frac{\partial}{\partial x^j} \{\sqrt{(-g)}F^{ij}\} \tag{56.3}$$

by equation (42.5) (g has been replaced by $-g$, since g is always negative for a real gravitational field). Equation (56.1) is accordingly equivalent to

$$\frac{1}{\sqrt{(-g)}} \frac{\partial}{\partial x^j} \{\sqrt{(-g)}F^{ij}\} = \mu_0 J^i \tag{56.4}$$

Also, in view of the skew-symmetry of the field tensor, it follows that equation (56.2) is equivalent to

$$\frac{\partial F_{ij}}{\partial x^k} + \frac{\partial F_{jk}}{\partial x^i} + \frac{\partial F_{ki}}{\partial x^j} = 0 \tag{56.5}$$

The energy–momentum tensor for the field is found from equation (29.5) to be given by

$$\mu_0 S^i_j = F^{ik} F_{jk} - \tfrac{1}{4}\delta^i_j F^{kl} F_{kl} \tag{56.6}$$

It now follows from Maxwell's equations that

$$S^j_{i;j} = -F_{ij} J^j = -D_i \tag{56.7}$$

where D_i is the 4-force density acting upon the charge distribution.

57. Black holes

The Schwarzschild metric is only valid in the region outside a spherically symmetric attracting body. Thus, if the radius of this body exceeds $2m$, the circumstance that the component g_{11} of the metric tensor becomes infinite at $r = 2m$ creates no difficulty. If, however, the body's radius is less than $2m$, the sphere $r = 2m$ lies in empty space and the nature of the field in the vicinity of this sphere needs careful study.

Although the metric is clearly invalid over the sphere $r = 2m$, it is an acceptable solution of the Einstein equation in the region $0 < r < 2m$. Consider a body moving radially in this region, not necessarily in a state of free fall. Then θ and ϕ

will both be constant and the metric equation reduces to

$$c^2 d\tau^2 = \alpha^{-1} (dr^2 - c^2 \alpha^2 dt^2) \tag{57.1}$$

where $\alpha = 2m/r - 1 > 0$, along the body's world-line. Given the equation of motion $r = r(t)$, this equation determines the proper time τ shown on a standard clock moving with the body. But $d\tau$ must be real and it follows that either (i) $dr/dt > c\alpha$ or (ii) $dr/dt < -c\alpha$. These inequalities show that it is impossible for a body to be stationary relative to our coordinate frame in this region. This implies that our picture of the frame as a set of coordinate clocks measuring the time t and stationary at the points (r, θ, ϕ) ceases to be applicable. Evidently, the static conditions we have been envisaging in the neighbourhood of the attracting body are not present in this region.

Next, consider a body falling freely along a radius towards the centre of attraction in the region $r > 2m$. Taking as initial conditions $t = 0$, $r = R$, $dr/dt = 0$, equations (53.5) and (53.6) lead to the equation of motion

$$\left(\frac{dr}{dt}\right)^2 = 2mc^2 \left(1 - \frac{2m}{R}\right)^{-1} \left(1 - \frac{2m}{r}\right)^2 \left(\frac{1}{r} - \frac{1}{R}\right) \tag{57.2}$$

Thus
$$ct = \left(\frac{R}{2m} - 1\right)^{1/2} \int_r^R \frac{r^{3/2} \, dr}{(r - 2m)(R - r)^{1/2}} \tag{57.3}$$

and it is clear that this integral diverges to $+\infty$ as $r \to 2m$. This means that, in the Schwarzschild frame, the body will need an infinite coordinate time to reach the sphere $r = 2m$. If the body is observed optically by an observer stationed at a considerable distance from the centre of attraction, since allowance must be made for the coordinate time needed for photons leaving the body to reach his telescope, the observed motion of the body, as measured by his coordinate clock, will be further retarded. But his coordinate clock will be almost indistinguishable from a standard clock and it follows that the apparent time of fall of the body to the Schwarzschild radius according to an external observer using an atomic clock will also be infinite.

If, however, instead of eliminating the proper time τ between equations (53.5) and (53.6), the coordinate time t is eliminated, the resulting equation is

$$\left(\frac{dr}{d\tau}\right)^2 = 2mc^2 \left(\frac{1}{r} - \frac{1}{R}\right) \tag{57.4}$$

After integration with $\tau = 0$ at $r = R$, this gives

$$c\tau = \sqrt{(R^3/2m)} \left[\sqrt{(\rho - \rho^2)} + \tfrac{1}{2} \cos^{-1} (2\rho - 1) \right] \tag{57.5}$$

where $\rho = r/R$ and the inverse cosine is taken in the first or second quadrants. τ will be the time recorded by a clock moving with the body and equation (57.5) shows that this remains finite for values of r through the value $2m$ to zero.

It is now evident that the reference frame we have been using is unacceptable if motions across the Schwarzschild sphere are to be studied and that, in particular, the coordinate time t becomes infinite at $r = 2m$ for some events which can occur in the experience of certain observers. It appears, therefore, that it is a deficiency in the reference frame which is responsible for the anomaly in the metric and our expectation is that the infinity can be removed by transformation to a new frame. This view of the matter is supported by the fact that g is finite at $r = 2m$, indicating that there is no singularity of space–time in this region.

The suggestion arising from our calculations is that t should be replaced by a new coordinate time u defined by a transformation equation

$$u = t + f(r) \tag{57.6}$$

where $f(r)$ becomes negatively infinite at $r = 2m$ in such a way as to cancel the infinity which we have seen to arise in t for certain events taking place on $r = 2m$. Substituting

$$dt = du - f'(r) dr \tag{57.7}$$

in the Schwarzschild metric, this transforms to

$$ds^2 = F dr^2 + r^2 (d\theta^2 + \sin^2 \theta d\phi^2) + c^2 (1 - 2m/r)(2f' dr du - du^2) \tag{57.8}$$

where

$$F = \frac{r}{r - 2m} - \frac{c^2}{r}(r - 2m)f'^2 \tag{57.9}$$

We can now remove the infinity by choosing $f(r)$ such that

$$cf' = r/(r - 2m) \tag{57.10}$$

Thus, we take

$$cf(r) = r + 2m \log (r - 2m) \tag{57.11}$$

and the metric then assumes the form

$$ds^2 = r^2 (d\theta^2 + \sin^2 \theta d\phi^2) + 2c dr du - c^2 (1 - 2m/r) du^2 \tag{57.12}$$

This metric must clearly satisfy Einstein's equation *in vacuo*. However, the field in the new frame is no longer static in the sense assumed in section 51; the presence of a term involving the first power of du shows that the field is not symmetric with respect to the past and future, i.e. the sense of description of its trajectory by a freely falling particle cannot be reversed with impunity.

Let us study, once again, a body moving radially, but not necessarily falling freely. Along its world-line, we have

$$ds^2 = 2c dr du - c^2 (1 - 2m/r) du^2 \tag{57.13}$$

Since ds^2 must be negative for any possible motion,

$$\frac{dr}{du} < \tfrac{1}{2} c (1 - 2m/r) \tag{57.14}$$

If $r < 2m$, this implies that dr/du is negative and that the body must move towards O; in particular, it cannot remain stationary. Thus, this is a region of irresistible collapse towards the centre of attraction for all physical bodies. It will be noted that the transformation has eliminated the possibility of outwards radial motion which existed when the Schwarzschild form of the metric was taken; this possibility can be recovered by changing the sign of u.

Now consider the motion of a body falling freely along a radius from an initial state of rest $dr/du = 0$ at $r = R > 2m$. Since $g_{41} = c, g_{44} = -c^2(1 - 2m/r)$, equation (53.5) must be replaced by

$$\frac{d}{d\tau}\left(\frac{dr}{d\tau} - c(1 - 2m/r)\frac{du}{d\tau}\right) = 0 \qquad (57.15)$$

Together with the first integral

$$2c\frac{dr}{d\tau}\frac{du}{d\tau} - c^2(1 - 2m/r)\left(\frac{du}{d\tau}\right)^2 = -c^2 \qquad (57.16)$$

this leads to the following quadratic for dr/du:

$$\left(\frac{dr}{du}\right)^2 + 4mc\left(\frac{1}{r} - \frac{1}{R}\right)\frac{dr}{du} + 2mc^2\left(1 - \frac{2m}{r}\right)\left(\frac{1}{R} - \frac{1}{r}\right) = 0 \qquad (57.17)$$

The roots are

$$\frac{dr}{du} = 2mc\left(\frac{1}{R} - \frac{1}{r}\right) \pm c\sqrt{\left[2m\left(1 - \frac{2m}{R}\right)\left(\frac{1}{r} - \frac{1}{R}\right)\right]} \qquad (57.18)$$

If $r > 2m$, one root is positive and one is negative. However, r must decrease initially (otherwise the square root in (57.18) becomes imaginary) and so the negative root is taken. r then decreases steadily to $r = 0$, its passage through the Schwarzschild radius being unremarkable. Once inside the Schwarzschild sphere, as already proved, the possibility of escape from the attraction no longer exists.

The world-lines of photons moving radially are null geodesics governed by the equation

$$2cdrdu - c^2(1 - 2m/r)du^2 = 0 \qquad (57.19)$$

There are two families of such geodesics, viz.

$$\frac{du}{dr} = 0, \quad \text{and} \quad c\frac{du}{dr} = \frac{2r}{r - 2m} \qquad (57.20)$$

For the first family, equation (57.7) gives

$$c\frac{dt}{dr} = -\frac{r}{r - 2m} \qquad (57.21)$$

provided $r > 2m$. This corresponds to a photon moving towards the centre of attraction. For the second family, we find

$$c\frac{\mathrm{d}t}{\mathrm{d}r} = \frac{r}{r-2m} \tag{57.22}$$

in the same region; i.e. a photon moving away from the centre of attraction. A photon belonging to the first family crosses the Schwarzschild sphere and then falls into O. Inside this sphere, the photons can be separated into two classes: (i) those for which u is constant along a world-line – these photons could have their source outside the sphere; (ii) those for which the second of equations (57.20) is valid and, hence,

$$cu = 2r + 4m\log(2m-r) + \text{constant} \tag{57.23}$$

As $r \to 2m, u \to -\infty$ and these photons cannot have had an external source. Since $\mathrm{d}u/\mathrm{d}r < 0$, these photons also fall into O.

It is now clear that, in the field described by the metric (57.12), no photon or particle can cross the Schwarzschild sphere in the sense r increasing. On the other hand, any photon or particle which crosses the sphere in the reverse sense is absorbed and cannot return to the external world. The conditions inside the sphere are accordingly referred to as a *black hole*. It is thought possible by astrophysicists that some stars may have collapsed under their own gravitational attraction to a radius less than their Schwarzschild radius. In such a case, as explained above, further contraction would become irresistible and the star would collapse to a singular point having infinite density. Such a collapse would require an infinite time by terrestrial clocks so that, assuming the age of the cosmos to be finite, it might be objected that no such objects can yet have come into existence. However, the idea of a cosmos of present events, all happening simultaneously relative to some universal time scale, is quite foreign to relativity theory, so that the objection is meaningless. The hard fact is that the possibility of a spaceship falling into the black hole created by such an object, in a time which is finite measured by an on-board clock, is a real one. A few cases of objects which appear to be in the early stages of gravitational collapse have already been detected.

If the metric (57.12) is transformed by changing the sign of u, another metric satisfying Einstein's equation is generated. A similar analysis shows that this governs the field in the vicinity of a *white hole*, where matter and photons can only cross the Schwarzschild sphere in an outgoing sense. Thus, a white hole behaves as an irresistible source and a black hole as an irresistible sink. Being invariant under a sign reversal of t, the Schwarzschild metric permits a black and a white hole to exist together.

58. Gravitational waves

Throughout this section it is assumed that the gravitational field is weak and that the coordinates x^i are quasi-Minkowskian, as explained in section 48. Thus, the metric tensor is given by equation (48.3) and terms of second or higher degree in the h_{ij} or their derivatives will be neglected. We shall further suppose that the coordinate frame is harmonic (see Exercises 5, No. 50), so that the metric tensor satisfies the condition

$$g^{ij}\Gamma^k_{ij} = 0 \tag{58.1}$$

It can be proved that a transformation of the form $\bar{x}^i = x^i + \xi^i(x)$, where the functions ξ^i are small with the h_{ij}, can always be made so that the $\bar{x}$-frame is harmonic (see Exercises 6, No. 37); this means that the harmonic coordinates will also be quasi-Minkowskian.

To the first order, equation (58.1) reduces to

$$[ii, k] = h_{ik, i} - \tfrac{1}{2}h_{ii, k} = 0 \tag{58.2}$$

Differentiation leads to

$$h_{ik, ij} - \tfrac{1}{2}h_{ii, jk} = 0 \tag{58.3}$$

Exchanging indices j, k and adding the new equation to (58.3), we get

$$h_{ij, ik} + h_{ik, ij} - h_{ii, jk} = 0 \tag{58.4}$$

The Ricci tensor has already been calculated to the first order of approximation at equation (49.2). Using the last result, this gives

$$R_{jk} = \tfrac{1}{2}h_{jk, ii} \tag{58.5}$$

Also

$$R = R_{jj} = \tfrac{1}{2}h_{jj, ii} \tag{58.6}$$

Thus, Einstein's tensor is given by

$$R_{jk} - \tfrac{1}{2}g_{jk}R = \tfrac{1}{2}h_{jk, ii} - \tfrac{1}{4}\delta_{jk}h_{rr, ii} = \tfrac{1}{2}h'_{jk, ii} \tag{58.7}$$

where

$$h'_{jk} = h_{jk} - \tfrac{1}{2}\delta_{jk}h_{rr} \tag{58.8}$$

Einstein's equation of gravitation is now expressible in the form

$$\square^2 h'_{jk} = h'_{jk, ii} = -2\kappa T_{jk} \tag{58.9}$$

The harmonic condition (58.2) can also be written

$$h'_{ik, i} = 0 \tag{58.10}$$

In empty space, equation (58.9) reduces to

$$\square^2 h'_{jk} = \left(\frac{\partial^2}{\partial x^2} + \frac{\partial^2}{\partial y^2} + \frac{\partial^2}{\partial z^2} - \frac{1}{c^2}\frac{\partial^2}{\partial t^2}\right)h'_{jk} = 0 \tag{58.11}$$

which is the wave equation, showing that gravitational waves are propagated *in vacuo* with the velocity of light.

In the case of a plane wave, we can write

$$h'_{jk} = A_{jk} \exp(ik_i x_i) \qquad (58.12)$$

where it is understood that the real part of the complex exponential is to be taken. Equation (58.11) is satisfied provided

$$k_i k_i = 0 \qquad (58.13)$$

and the condition (58.10) also requires that

$$k_i A_{ik} = 0 \qquad (58.14)$$

Since A_{ik} is symmetric, the last equation shows that the A_{i4} can be expressed in terms of the $A_{\alpha\beta}$ ($\alpha, \beta = 1, 2, 3$). By further transformation of coordinates, it can be shown that all amplitudes can be expressed in terms of two parameters only and hence that gravitational waves have, essentially, only two modes of polarization.

The solution of the wave equation (58.9) with source term $-2\kappa T_{jk}$ is well known to be given by Kirchhoff's formula (Bateman, 1952):

$$h'_{jk}(\mathbf{x}_0, t_0) = \frac{\kappa}{2\pi} \int_V \frac{1}{r} T_{jk}(\mathbf{x}, t_0 - r/c) \, dV \qquad (58.15)$$

where $\mathbf{x}_0 = (x_0^1, x_0^2, x_0^3)$, $\mathbf{x} = (x^1, x^2, x^3)$ are position vectors with respect to the origin O of the frame in use, and $r = |\mathbf{x}_0 - \mathbf{x}|$ is the distance between these points; V is the region of space over which T_{jk} is non-vanishing. Note that t_0 is retarded in the integrand by a time r/c, since the effect of the source at $\mathbf{x}$ will not be felt at $\mathbf{x}_0$ until the time for its transmission over the distance r has elapsed. In the case of a source which is confined to a small region of space including the origin O, if $r_0 = |\mathbf{x}_0|$ is large compared with the dimensions of this region, equation (58.15) can be approximated by

$$h'_{jk}(\mathbf{x}_0, t_0) = \frac{\kappa}{2\pi r_0} \int_V T_{jk}(\mathbf{x}, t_0 - r_0/c) \, dV \qquad (58.16)$$

But, as explained above, the components h'_{j4} can be obtained easily once the $h'_{\alpha\beta}$ have been found. We shall now show that a further simplification of the last formula is possible in these cases.

First note that

$$(T_{\alpha\gamma} x^\beta)_{,\gamma} = T_{\alpha\gamma,\gamma} x^\beta + T_{\alpha\beta} \qquad (58.17)$$

Since the divergence of T_{ij} vanishes, we have

$$T_{\alpha\gamma,\gamma} + T_{\alpha4,4} = 0 \qquad (58.18)$$

and thus

$$(T_{\alpha\gamma} x^\beta)_{,\gamma} = T_{\alpha\beta} - T_{\alpha4,4} x^\beta \qquad (58.19)$$

Integrating over the region V, the integral of the left-hand member is seen to be zero by application of Gauss's divergence theorem (assuming $T_{\alpha\gamma}$ vanishes over the bounding surface); we accordingly obtain the result

$$\int T_{\alpha\beta} dV = \int T_{\alpha4,4} x^\beta dV = -\frac{i}{c} \frac{d}{dt} \int T_{\alpha4} x^\beta dV \qquad (58.20)$$

Exchanging the indices α, β and adding the new identity to (58.20), we find

$$\int T_{\alpha\beta} dV = -\frac{i}{2c} \frac{d}{dt} \int (T_{\alpha4} x^\beta + T_{\beta4} x^\alpha) dV \qquad (58.21)$$

We next integrate the identity

$$(T_{\gamma4} x^\alpha x^\beta)_{,\gamma} = T_{4\gamma,\gamma} x^\alpha x^\beta + T_{\alpha4} x^\beta + T_{\beta4} x^\alpha \qquad (58.22)$$

over V. By the divergence theorem, the integral of the left-hand member vanishes. Hence

$$\int (T_{\alpha4} x^\beta + T_{\beta4} x^\alpha) dV = -\int T_{4\gamma,\gamma} x^\alpha x^\beta dV \qquad (58.23)$$

Equations (58.21) and (58.23) now yield the result

$$\int T_{\alpha\beta} dV = \frac{i}{2c} \frac{d}{dt} \int T_{4\gamma,\gamma} x^\alpha x^\beta dV = -\frac{i}{2c} \frac{d}{dt} \int T_{44,4} x^\alpha x^\beta dV$$

$$= -\frac{1}{2c^2} \frac{d^2}{dt^2} \int T_{44} x^\alpha x^\beta dV \qquad (58.24)$$

where we have again made use of the equation $T_{ij,j} = 0$. But, equation (21.14) shows that $T_{44} = -\mu c^2$ and equations (58.16) and (58.24) therefore lead to the final result

$$h'_{\alpha\beta}(\mathbf{x}_0, t) = \frac{2G}{c^4 r_0} \frac{d^2}{dt^2} \int \mu(\mathbf{x}, t - r_0/c) x^\alpha x^\beta dV \qquad (58.25)$$

It should be noted that it is the second time derivative of the *second* moment of the mass distribution which is responsible for the gravitational wave. In the corresponding electromagnetic situation, it is the second time derivative of the *first* moment of the charge distribution which is responsible for the electromagnetic wave.

Instruments have been devised to detect the small variations in the gravitational field caused by waves proceeding from possible sources within the galaxy (e.g. pulsating neutron stars, binary stars or supernova explosions), but no clearly unambiguous results have yet been obtained. Such instruments attempt to measure the small strains induced in very large masses of metal by the tidal forces caused by the passage through them of gravitational waves.

Exercises 6

1. If the y-frame is defined as in section 47, show that the metric tensor in the x-frame is given by

$$g_{ij} = \frac{\partial y^k}{\partial x^i} \frac{\partial y^k}{\partial x^j}$$

Hence lower the index j in T^{ij} defined by equation (47.2) and show that the result is T^i_j as defined in equation (47.3).

2. Given that space–time has the metric

$$ds^2 = dx^2 + dy^2 + e^{2\theta}dz^2 - e^{2\phi}dt^2$$

where θ, ϕ are functions of z only, prove that the Riemann–Christoffel tensor vanishes if, and only if,

$$\phi'' - \theta'\phi' + \phi'^2 = 0$$

where dashes denote differentiations with respect to z. If $\phi = -\theta$, prove that the space–time is flat provided $\phi = \frac{1}{2}\log(a + bz)$, where a, b are constants.

3. If space–time has the metric

$$ds^2 = e^\lambda(dr^2 + dz^2) + r^2 e^{-\rho}d\phi^2 - e^\rho dt^2$$

where λ, ρ are functions of r and z only, show that the field equations in empty space $R_{ij} = 0$ require that λ and ρ should satisfy the equations

$$\lambda_1 + \rho_1 = \frac{1}{2}r(\rho_1^2 - \rho_2^2)$$
$$\lambda_2 + \rho_2 = r\rho_1\rho_2$$
$$\rho_{11} + \rho_{22} + \frac{1}{r}\rho_1 = 0$$
$$\lambda_{11} + \lambda_{22} + \rho_{11} + \rho_{22} + \frac{1}{2}(\rho_1^2 + \rho_2^2) = 0$$

where subscripts 1 and 2 denote partial differentiations with respect to r and z respectively.

4. If space–time has the metric

$$ds^2 = e^{2kx}(dx^2 + dy^2 + dz^2 - dt^2)$$

where k is constant, and $v^2 = \dot{x}^2 + \dot{y}^2 + \dot{z}^2$, dots denoting differentiations with respect to t, show that for a freely falling body

$$1 - v^2 = (1 - V^2)e^{2kx}$$

where $v = V$ at $x = 0$.

5. If space–time has the metric

$$ds^2 = \alpha^2(dx^2 + dy^2 + dz^2) - c^2\alpha dt^2$$

where $\alpha = 1/(1 - kx)$ and k is constant, and v is as defined in the previous exercise, prove that for a freely falling body

$$V^2 - v^2 = kc^2 x$$

where $v = V$ at $x = 0$.

6. The space–time metric over a certain region of empty space is

$$ds^2 = e^\alpha (dx^2 + dy^2 + dz^2) - e^\beta dt^2$$

where α, β are functions of z alone. Show that Einstein's equation is satisfied provided

$$\alpha'' + \tfrac{1}{2}\alpha'^2 + \tfrac{1}{2}\alpha'\beta' = 0$$
$$\alpha'' + \tfrac{1}{2}\beta'' + \tfrac{1}{4}\beta'^2 - \tfrac{1}{4}\alpha'\beta' = 0$$
$$\beta'' + \tfrac{1}{2}\beta'^2 + \tfrac{1}{2}\alpha'\beta' = 0$$

Deduce that $e^\alpha = A(k - z)^4$, $e^\beta = B(k - z)^{-2}$, where A, B, k are constants.

7. Show that the space–time metric

$$ds^2 = e^\alpha dr^2 + r^2 d\theta^2 + e^\beta dz^2 - e^\gamma dt^2$$

where r, θ, z are quasi-cylindrical polar coordinates and t is the time and α, β, γ are functions of r alone, satisfies Einstein's equation *in vacuo*, provided

$$\beta'' + \gamma'' + \tfrac{1}{2}\beta'^2 + \tfrac{1}{2}\gamma'^2 - \tfrac{1}{r}\alpha' - \tfrac{1}{2}\alpha'\beta' - \tfrac{1}{2}\alpha'\gamma' = 0$$
$$\alpha' = \beta' + \gamma'$$
$$\beta'' + \tfrac{1}{2}\beta'^2 - \tfrac{1}{2}\alpha'\beta' + \tfrac{1}{2}\beta'\gamma' + \tfrac{1}{r}\beta' = 0$$
$$\gamma'' + \tfrac{1}{2}\gamma'^2 - \tfrac{1}{2}\alpha'\gamma' + \tfrac{1}{2}\beta'\gamma' + \tfrac{1}{r}\gamma' = 0$$

dashes indicating differentiations with respect to r. Deduce that

$$e^\alpha = Ar^{-(\lambda + \mu)}, \ e^\beta = Br^{-\lambda}, \ e^\gamma = Cr^{-\mu}$$

where A, B, C, λ, μ are constants and $\lambda\mu = 2(\lambda + \mu)$.

8. A certain region of space–time has metric

$$ds^2 = dx^2 + dy^2 + dz^2 - x^2 dt^2$$

A particle is stationary at the point $x = 1$, $y = z = 0$ at $t = 0$. If the particle is released at this instant and falls freely, show that it moves along the x-axis with equation of motion $x = \operatorname{sech} t$. A photon is emitted from the point $(1, 0, 0)$ at $t = 0$ in the direction of the positive y-axis. Show that at this instant $\dot{x} = \dot{z} = 0$, $\dot{y} = 1$ and that the path of the photon is the circle $x^2 + y^2 = 1$.

9. De Sitter's universe has metric

$$ds^2 = A^{-1} dr^2 + r^2 (d\theta^2 + \sin^2 \theta d\phi^2) - Ac^2 dt^2$$

where $A = 1 - r^2/R^2$, R being constant. At $t = 0$, a photon leaves the origin $r = 0$ and travels outwards along the straight line $\theta = $ constant, $\phi = $ constant. Find its coordinate r at time t and show that $r = \tfrac{1}{2}R$ when $t = R(\log 3)/2c$ and that $r \to R$ as $t \to \infty$.

10. r, θ, z are quasi-cylindrical coordinates in a gravitational field determined by the metric

$$ds^2 = r^2(dr^2 + d\theta^2) + r(dz^2 - dt^2)$$

A particle is projected from the point $r = 1, \theta = 0, z = 0$ in the field with such velocity that $\dot{r} = \dot{z} = 0, \dot{\theta} = \sqrt{3}/2$ (dots denote differentiations with respect to t). Prove that, if the particle falls freely, it moves in the plane $z = 0$ between the circles $r = 1, r = 3$, first touching the outer circle where $\theta = \sqrt{3}\pi$. A photon is emitted from the point $r = 1, \theta = 0, z = 0$ and moves initially so that $\dot{r} = \dot{z} = 0$. Prove that its path is the spiral $r = 1 + \frac{1}{4}\theta^2$ in the plane $z = 0$.

11. The metric for de Sitter's universe can be expressed in the form

$$ds^2 = e^{2ct/R}(dx^2 + dy^2 + dz^2) - c^2 dt^2$$

where R is a constant and x, y, z can be treated as rectangular Cartesian coordinates. Show that the trajectories of freely falling particles and photons are straight lines. A particle is projected from the origin at $t = 0$ with a velocity V along the positive x-axis. Prove that its x-coordinate at time t is given by

$$Vx = R\left[c - \sqrt{(c^2 - V^2 + V^2 e^{-2ct/R})}\right]$$

A body at the point $x = X$ on the x-axis emits a photon towards the origin at $t = 0$. Show that the photon arrives at O at time $t = -\dfrac{R}{c}\log(1 - X/R)$. Discuss the case where $X > R$.

12. r, θ, ϕ are quasi-spherical polar coordinates in a gravitational field which is spherically symmetric about a centre of attraction $r = 0$. The space–time metric is

$$ds^2 = \left(\frac{r}{r+1}\right)^2 dr^2 + r^2 d\theta^2 + r^2 \sin^2\theta d\phi^2 - \frac{r}{r+2}dt^2$$

A particle is projected from the point $r = 1, \theta = \frac{1}{2}\pi, \phi = 0$ at $t = 0$ with velocity such that $\dot{r} = 0, \dot{\theta} = 0, \dot{\phi} = 1/\sqrt{6}$ and falls freely. Show that the particle's trajectory lies in the plane $\theta = \frac{1}{2}\pi$ and has polar equation

$$r = \frac{5 - \cos(a\phi)}{3 + \cos(a\phi)}$$

where $a = \sqrt{(8/3)}$. Deduce that the particle moves between two circles of radii 1 and 3 and calculate the increment in ϕ between two successive contacts with one of these circles. (Ans. $2\pi/a$.)

13. Taking the metric for de Sitter's universe in the form stated in exercise 9, find equations of motion for a particle projected from the point $r = \frac{1}{2}R$, $\theta = \frac{1}{2}\pi, \phi = 0$ with such velocity that $\dot{r} = \dot{\theta} = 0, \dot{\phi} = \sqrt{\dfrac{3}{2}\dfrac{c}{R}}$ and thereafter falls under gravity. Show that its trajectory lies in the plane $\theta = \frac{1}{2}\pi$ and that its polar

equation is

$$r = R/(5\cos^2\phi - 1)^{1/2}.$$

14. $Oxyz$ is a quasi-rectangular Cartesian coordinate frame constructed in a certain gravitational field. If t is the time measured by a system of clocks stationary in the frame, the space–time metric is $ds^2 = z(dx^2 + dy^2 + dz^2 - dt^2)$. A particle is projected from the point $(0, 0, 1)$ at $t = 0$ with velocity components $\dot{x} = v (< 1)$, $\dot{y} = \dot{z} = 0$ and thereafter falls freely. Show that its trajectory is a parabola and lies in the xz-plane and that the particle arrives at the xy-plane at time $t = 2/\sqrt{(1 - v^2)}$.

15. Show that Einstein's equation (47.16) can be written in the form

$$R^{ij} + \Lambda g^{ij} = \kappa(\tfrac{1}{2}Tg^{ij} - T^{ij})$$

16. Inside a static gravitating homogeneous sphere of liquid, the proper density is μ (a constant) and the pressure is p. The energy–momentum tensor has zero components except for $T_1^1 = T_2^2 = T_3^3 = p$, $T_4^4 = -c^2\mu$. Assuming that the metric of the field inside the sphere is given by equation (51.10) with $a = e^\alpha$ and $b = e^\beta$, show that Einstein's equation (47.15) can be satisfied by making α, β and p satisfy the equations

$$\frac{d}{dr}\{r(1 - e^{-\alpha})\} = \kappa c^2 \mu r^2$$

$$\beta' = \frac{1}{r}(e^\alpha - 1) + \kappa r e^\alpha p$$

$$\beta'' + \tfrac{1}{2}\beta'^2 - \tfrac{1}{2}\alpha'\beta' - \frac{2}{r}\alpha' = \kappa e^\alpha(p - c^2\mu)$$

dashes denoting differentiations with respect to r. Assuming $\alpha = 0$ at $r = 0$ and $p = 0$ at $r = a$ (the surface), deduce that

$$e^{-\alpha} = 1 - qr^2$$

where $q = \kappa c^2 \mu/3$, and that

$$p = c^2\mu \frac{(1 - qr^2)^{1/2} - (1 - qa^2)^{1/2}}{3(1 - qa^2)^{1/2} - (1 - qr^2)^{1/2}}$$

17. Obtain the equations of motion of photons moving radially inside the Schwarzschild sphere and deduce that a photon moving away from the centre O takes an infinite coordinate time t to reach the sphere and a photon moving towards the centre from $r = R (< 2m)$ takes a time $t = T$ given by

$$cT = -R - 2m\log(1 - R/2m)$$

to reach O.

18. Obtain equations (57.2) and (57.3) for a body falling freely towards the

centre of attraction in the region $r > 2m$ and hence prove that

$$ct = \left(\frac{R}{2m} - 1\right)^{1/2} \left[\sqrt{\{r(R-r)\}} + (R + 4m)\cos^{-1}(r/R)^{1/2} \right] - 2m\log\left(\frac{1-\gamma}{1+\gamma}\right)$$

where

$$\gamma = \left[\frac{2m(R-r)}{r(R-2m)}\right]^{1/2}.$$

Deduce that $t \to \infty$ as $r \to 2m$.

19. Obtain equations (57.4) and (57.5) and deduce that the time recorded on a standard clock attached to a freely falling body as it falls from the Schwarzschild sphere to the centre of attraction is $\pi m/c$. Calculate this time in the case of a black hole having solar mass. (Ans. 16 μs.)

20. Verify that the metric tensor given by equation (57.12) satisfies Einstein's equation in empty space.

21. Show that the Kruskal–Szekeres transformation

$$u = (r/2m - 1)^{1/2} e^{r/4m} \cosh(ct/4m)$$
$$v = (r/2m - 1)^{1/2} e^{r/4m} \sinh(ct/4m)$$

converts the Schwarzschild metric to the form

$$ds^2 = \frac{32m^3}{r} e^{-r/2m}(du^2 - dv^2) + r^2(d\theta^2 + \sin^2\theta d\phi^2)$$

where r is given in terms of u, v by the equation

$$u^2 - v^2 = (r/2m - 1)e^{r/2m}$$

Deduce that the world-lines of radially moving photons are $u \pm v = $ constant.

22. Show that the transformation

$$u = v + \frac{2}{3a}r^{3/2}$$
$$v = t + 2ar^{1/2} - a^2\log\frac{r^{1/2} + a}{r^{1/2} - a},$$

where $a^2 = 2m$, puts the Schwarzschild metric into the form

$$ds^2 = \frac{4}{9}\mu^2(u - v)^{-2/3}du^2 + \mu^2(u - v)^{4/3}(d\theta^2 + \sin^2\theta d\phi^2) - dv^2$$

where $\mu^3 = 9a^2/4$.

23. A photon is emitted from the point $r = m$, $\theta = \frac{1}{2}\pi$, $\phi = 0$, inside a black hole (Schwarzschild coordinates) with angular velocities $\dot\theta = 0$, $\dot\phi = 3\sqrt{3}c/m$. Show that $\dot r = \pm 2\sqrt{7}c$ initially. In the case when the initial value of $\dot r$ is negative,

show that the photon moves in the plane $\theta = \frac{1}{2}\pi$ and falls into the centre along the trajectory

$$\frac{6m}{r} = 3\coth^2\tfrac{1}{2}(\alpha - \phi) - 1$$

where $\alpha = \log\frac{1}{2}(5 + \sqrt{21})$.

24. Show that the only possible circular orbits for a photon in a Schwarzschild field all have radius $r = 3m$ and that their period in coordinate time is $6\sqrt{3}\pi m/c$. Show that these orbits are unstable.

25 A body moves in a circular orbit of radius r in the plane $\theta = \frac{1}{2}\pi$ in a Schwarzschild field. Show that $r > 3m$ and that the angular velocity $d\phi/dt$ is related to r in the same way as in classical theory. Show that the period of the motion as measured by a standard clock attached to the body is

$$\frac{2\pi r}{c}\left(\frac{r}{m} - 3\right)^{\frac{1}{2}}$$

Show, also, that the period as measured by an observer using a standard clock who is stationary at some point on the orbit is

$$\frac{2\pi r}{c}\left(\frac{r}{m} - 2\right)^{\frac{1}{2}}$$

Show that the orbit is unstable if $3m < r < 6m$, but is stable otherwise.

26. r, θ, ϕ are Schwarzschild coordinates. A fixed observer at the point R, θ, ϕ transmits a wireless signal radially towards the attracting body. The signal is reflected by a small body at the point r, θ, ϕ and returns to the observer. Show that the time elapsing between transmission and reception as measured by the observer's standard clock is

$$\frac{2}{c}(1 - 2m/R)^{1/2}\left(R - r + 2m\log\frac{R - 2m}{r - 2m}\right)$$

Calculate the distance covered by the signal and deduce that, according to classical theory, the time for the double journey would be

$$\frac{2}{c}\left[(R^2 - 2mR)^{1/2} - (r^2 - 2mr)^{1/2} + 2m\log\frac{R^{1/2} + (R - 2m)^{1/2}}{r^{1/2} + (r - 2m)^{1/2}}\right]$$

Show that, to the first order in m, the difference between these times is

$$\frac{2m}{c}\left(\log\frac{R}{r} + \frac{r}{R} - 1\right)$$

(*Note*: This result suggests a method of checking the general theory using the Sun's field and Mercury or Venus as the reflector.)

27. An atom, which is stationary at a Schwarzschild coordinate distance r from the centre of a spherically symmetric body, emits light of frequency v which is observed by a stationary observer at a coordinate distance R ($> r$) from the centre. Show that the observed frequency is $v - \delta v$, where

$$\delta v / v = m\left(\frac{1}{r} - \frac{1}{R}\right)$$

to the first order in m.

28. By replacing the spherical polar coordinate r occurring in the Schwarzschild metric (52.10) by a new coordinate r' where

$$r = r'\left(1 + \frac{m}{2r'}\right)^2$$

obtain this metric in 'isotropic' form, viz.

$$ds^2 = \left(1 + \frac{m}{2r'}\right)^4 (dr'^2 + r'^2 d\theta^2 + r'^2 \sin^2\theta \, d\phi^2) - \left(\frac{1 - m/2r'}{1 + m/2r'}\right)^2 c^2 dt^2$$

29. Employing a certain frame, an event is specified by spatial coordinates (x, y, z) and a time t. The corresponding space–time manifold has metric

$$ds^2 = dx^2 + dy^2 + dz^2 + 2at \, dx \, dt - (c^2 - a^2 t^2) dt^2$$

Show that a particle falling freely in the gravitational field observed in the frame has equations of motion

$$x = A + Bt - \tfrac{1}{2}at^2, \quad y = C + Dt, \quad z = E + Ft,$$

where A, B, C, D, E, F are constants. By transforming to coordinates (x', y, z, t), where $x' = x + \tfrac{1}{2}at^2$, and recalculating the metric, explain this result.

30. (x^1, x^2, x^3) are spatial coordinates of an event relative to a frame S and x^4 is the time of the event measured by a clock in S. A second frame I is falling freely in the neighbourhood of the event and may be regarded as inertial. $Oy^1 y^2 y^3$ are rectangular Cartesian axes in I and y^4/ic represents the time within I as measured by synchronized clocks attached to the frame. Show that g_{ij}, the metric tensor in S, is given by

$$g_{ij} = \frac{\partial y^k}{\partial x^i} \frac{\partial y^k}{\partial x^j}.$$

P is a point, fixed in S, having coordinates (x^1, x^2, x^3). At the instant x^4, I is chosen so that P is instantaneously at rest in I. Deduce that

$$\frac{\partial y^4}{\partial x^i} = \frac{g_{i4}}{\sqrt{(g_{44})}}$$

at $x^i . dl$ is the distance between P and a neighbouring point

$$P'(x^1 + dx^1, x^2 + dx^2, x^3 + dx^3)$$

as measured by a standard rod in I at the instant x^4. Prove that

$$dl^2 = dy^\alpha dy^\alpha = \gamma_{\lambda\mu} dx^\lambda dx^\mu$$

where α, λ, μ range over the values 1, 2, 3, and

$$\gamma_{\lambda\mu} = g_{\lambda\mu} - \frac{g_{\lambda 4} g_{\mu 4}}{g_{44}}$$

($\gamma_{\lambda\mu}$ is the metric tensor for the $\mathcal{R}_3$ which is S at the instant x_4.)

31. $Oxyz$ is a rectangular Cartesian inertial frame I. A rigid disc rotates in the xy-plane about its centre O with angular velocity ω. Polar coordinates (r, θ) in a frame R rotating with the disc are defined by the equations

$$x = r\cos(\theta + \omega t), \quad y = r\sin(\theta + \omega t)$$

where t is the time measured by synchronized clocks in the inertial frame. If the time of an event in R is taken to be the time shown by an adjacent clock in I, show that the space–time metric associated with R is

$$ds^2 = dr^2 + r^2 d\theta^2 + 2\omega r^2 d\theta dt - (c^2 - r^2\omega^2) dt^2$$

Deduce that the metric for geometry in R is given by

$$dl^2 = dr^2 + \frac{r^2 d\theta^2}{1 - \omega^2 r^2/c^2}$$

(*Hint*: employ the result of the previous exercise.) Hence show that the family of geodesics on the disc is determined by the equation

$$\theta = \text{const.} - \sin^{-1}\left(\frac{a}{r}\right) - \frac{a}{r^2}\sqrt{(r^2 - a^2)}$$

where $r_1 = c/\omega$ and $|a| < r_1$. Sketch this family. What is the physical significance of r_1?

32. $x^i (i = 1, 2, 3, 4)$ are three space coordinates and time relative to a reference frame S. A test particle is momentarily at rest in S at the point (x^1, x^2, x^3) at the time x^4. If g_{ij} is the metric tensor for the gravitational field in S, write down the conditions that the world-line of the particle is a geodesic and deduce that

$$g_{i\alpha} \frac{d^2 x^\alpha}{(dx^4)^2} = \frac{1}{2}\left(\frac{\partial g_{44}}{\partial x^i} + \frac{g_{i4}}{g_{44}}\frac{\partial g_{44}}{\partial x^4}\right) - \frac{\partial g_{i4}}{\partial x^4}$$

at the point x^i. Hence show that the covariant components of the particle's acceleration in S are given by

$$\gamma_{\alpha\beta}\frac{d^2 x^\beta}{(dx^4)^2} = -\frac{\partial U}{\partial x^\alpha} - (c^2 + 2U)^{1/2}\frac{\partial \gamma_\alpha}{\partial x^4}$$

where $\gamma_{\alpha\beta}$ is defined in exercise 30 and

$$g_{44} = -(c^2 + 2U), \quad \gamma_\alpha = g_{\alpha 4}/\sqrt{(-g_{44})}$$

(U, γ_α are the gravitational scalar and vector potentials respectively.)

Show that, in the case of the space–time metric appropriate to the rotating frame of exercise 31, the gravitational vector potential vanishes and the scalar potential is given by $U = \frac{1}{2}\omega^2 r^2$. Interpret this result in terms of the centrifugal force.

33. De Sitter's universe has metric

$$ds^2 = A^{-1}dr^2 + r^2d\theta^2 + r^2\sin^2\theta d\phi^2 - Ac^2dt^2$$

where $A = 1 - r^2/R^2$, R being constant. Obtain the differential equations satisfied by the null geodesics and show that along null geodesics in the plane $\theta = \frac{1}{2}\pi$,

$$a\frac{dr}{d\phi} = r(r^2 - a^2)^{1/2}$$

where a is a constant. Deduce that, if r, ϕ are taken to be polar coordinates in this plane, the paths of light rays in this universe are straight lines.

34. Einstein's universe has the metric

$$ds^2 = \frac{1}{1 - \lambda r^2}dr^2 + r^2d\theta^2 + r^2\sin^2\theta d\phi^2 - c^2dt^2$$

where (r, θ, ϕ) are spherical polar coordinates. Obtain the equations governing the null geodesics and show that, in the plane $\theta = \frac{1}{2}\pi$, these curves satisfy the equation

$$\left(\frac{dr}{d\phi}\right)^2 = r^2(1 - \lambda r^2)(\mu r^2 - 1)$$

where μ is a constant. Putting $r^2 = 1/v$, integrate this equation and hence deduce that the paths of light rays in the plane $\theta = \frac{1}{2}\pi$ are the ellipses

$$\lambda x^2 + \mu y^2 = 1$$

where (x, y) are rectangular Cartesian coordinates. Show, also, that the time taken by a photon to make one complete circuit of an ellipse is $2\pi/(c\lambda^{1/2})$.

35. If the metric of space–time is

$$ds^2 = \alpha^2(dx^2 + dy^2 + dz^2) - k\alpha dt^2$$

where α is a function of x alone and k is a constant, obtain the differential equations governing the world-lines of freely falling particles. If x, y, z are interpreted as rectangular Cartesian coordinates by an observer and t is his time variable, show that there is an energy equation for the particles in the form

$$\frac{1}{2}v^2 - \frac{k}{2\alpha} = \text{constant}$$

36. (r, θ, ϕ, t) are interpreted as spherical polar coordinates and time. A gravitational field is caused by a point electric charge at the pole. Assuming that

the space–time metric is given by equation (51.10) and that the 4-vector potential for the electromagnetic field of the charge is given by $\Omega_i = (0, 0, 0, \chi)$, where $\chi = \chi(r)$, calculate the covariant components of the field tensor F_{ij} from equation (26.6) and deduce the contravariant components F^{ij}. Assuming that $J^i = 0$, prove that Maxwell's equations (56.4) and (56.5) are all satisfied if

$$\frac{d\chi}{dr} = \frac{q}{4\pi\varepsilon_0 r^2} \sqrt{(ab)}$$

where q is a constant.

Calculate the elements of the mixed energy–momentum tensor from equation (56.6) and write down Einstein's equations (47.15) for the gravitational field. Show that these are satisfied provided

$$\frac{1}{a} = b = 1 - \frac{2m}{r} + \frac{q^2 G}{4\pi\varepsilon_0 c^4} \cdot \frac{1}{r^2}$$

where m is a constant.

37. If the coordinates x^i are quasi-Minkowskian so that the metric tensor is given by equation (48.3), show that the transformation $\bar{x}^i = x^i + \xi^i(x)$ makes the $\bar{x}$-frame harmonic provided the ξ^i satisfy the conditions

$$\xi^i{}_{,jj} = h_{ij,j} - \tfrac{1}{2} h_{jj,i}$$

(Neglect second order terms in the h_{ij} and use the condition given in Exercises 5, No. 50.) Show also that $\bar{h}_{ii} = 0$ provided the functions ξ^i satisfy the additional condition $\xi^i{}_{,i} = \tfrac{1}{2} h_{ii}$. If the x-frame is harmonic before transformation, show that the $\bar{x}$-frame is also harmonic provided $\xi^i{}_{,jj} = 0$.

38. A sphere of mass M is expanding in such a manner that its density remains uniform. If $a(t)$ is its radius at time t, show that, at a large distance r from its centre, the gravitational wave generated has components

$$h'_{11} = h'_{22} = h'_{33} = \frac{4GM}{5rc^4} (\dot{a}^2 + a\ddot{a})$$

the components $h'_{12}, h'_{23}, h'_{31}$ being zero. (The bracketed expression is to be calculated at the appropriate retarded time.)

39. A uniform rod of mass M and length $2a$ is pivoted with its centre at the origin of the x-frame and rotates in the $x_2 x_3$-plane with angular velocity ω. Show that, at a large distance r from the rod, the gravitational wave generated has non-zero components

$$h'_{22} = -h'_{33} = A \cos 2\omega t, \qquad h'_{23} = A \sin 2\omega t$$

where $A = 4GMa^2\omega^2/3rc^4$ and the instant $t = 0$ has been chosen appropriately.

40. Show that, if the cosmical constant term is retained in Einstein's equation, it reduces in empty space to $R_{ij} + \Lambda g_{ij} = 0$. Deduce that the spherically symmetric

Schwarzschild solution (cf. equation (52.9)) is given by

$$b = 1 - \frac{2m}{r} - \frac{1}{3}\Lambda r^2$$

Using the approximate equation (48.17), show that this implies the existence of an additional force of repulsion from the centre proportional to the radius r.

CHAPTER 7

Cosmology

59. Cosmological principle. Cosmical time

Cosmology is the study of the large-scale features of the universe, such as the distribution and motions of the galaxies and the density of radiation and dust through intergalactic space. It is also concerned with the manner in which these features can be expected to change over very long periods of time measured in billions (10^9) of years, i.e. with the evolution of the cosmos. Such calculations also throw light on the stages through which the cosmos has passed to arrive at its present state, and attempt to answer the question, did the universe have a beginning in time or has it always existed? If the universe had a beginning, as the evidence now strongly suggests, the study of its state during the very early stages of its evolution is called *cosmogony*.

Since the galaxies are electrically uncharged, the only force influencing their motion is gravity. Thus, cosmology is necessarily founded on a theory of gravitation. It has been shown (see, e.g., Bondi, 1960) that the Newtonian theory is quite capable of generating models for the cosmos which provide explanations for many of its observed features. However, these models necessarily assume that space is Euclidean, whereas Einstein's theory indicates that, in the presence of a gravitational field, space becomes curved and its geometry is then Riemannian. The curvature generated by the gravitational attraction of a galaxy is inappreciable and may be disregarded so long as we confine our attention to regions of space whose dimensions are comparable with those of a galaxy, but this effect has major consequences when the spatial extension of the whole cosmos is considered; in particular, as we shall see, a possible consequence is that the total volume of space is finite and, therefore, that the universe is not potentially infinite in extent as a Newtonian cosmology must assume. Only cosmic models which are in accord with general relativity theory will accordingly be studied.

The reader will be presumed familiar with the basic facts relating to the distribution of matter and radiation over the cosmos as it is observed in the present epoch (Rowan-Robinson, 1979). The mass of the radiation is roughly one-thousandth of the mass of the galactic matter and its gravitational effect is therefore negligible by comparison with the attraction of the galaxies. However, at earlier epochs, the contribution to the gravitational field of the radiation was probably much more considerable and, during the first million years after the 'big bang', it is thought that the cosmos was dominated by its radiation; this radiation is assumed to have been in equilibrium with the matter and hence to have acquired a black-body frequency distribution. The remnant of this black-body radiation in the present epoch was detected by Penzias and Wilson in 1965 and this still forms the major part of the total cosmic radiation. It is not known what proportion of the matter in the universe has been attracted into the galaxies; the density of matter in intergalactic space is certainly so small as to have no observable effect on the light transmitted through these regions from the most distant sources, but the volume of space is so large that this observation is not inconsistent with the hypothesis that the net mass of intergalactic matter is many times that of the matter present in the galaxies. As will be seen, our ignorance in regard to this datum prevents our reaching a firm conclusion whether the cosmos is finite or infinite in extent. Although the galaxies often occur in clusters, from our viewpoint their overall distribution appears to be isotropic and homogeneous. At very great distances, the galactic density is observed to increase, but it must be remembered that such observations are carried out by light which was emitted at a much earlier epoch when all the galaxies are thought to have been closer together; it is assumed that, at the 'present cosmical time' (precise definition follows later), the density of galactic mass is uniform throughout the cosmos.

That the galactic density is decreasing as the universe evolves is in accordance with the observed recession of the galaxies. To be more precise, what is observed is that the spectrum of the light from a distant galaxy is shifted towards the red end of the spectrum by an amount which is approximately proportional to its distance. This is *Hubble's law*. The reduction in frequency is interpreted as a Doppler effect caused by the motion of the galaxy away from the observer along the line of sight. Since it is supposed that the whole universe is in a state of expansion, each galactic observer will experience a recession of all the other galaxies in accordance with Hubble's law. Clearly, if matter is conserved during this expansion, a steady reduction in its density is inevitable and there is now an accumulation of evidence that the matter density was indeed greater in the distant past than it is today. However, a steady-state cosmology (Bondi and Gold, 1948, Hoyle, 1948) has been proposed in which the galactic density remains constant due to the continuous creation of matter in intergalactic space; this matter condenses into new galaxies and so maintains a steady-state distribution.

As a first step towards the construction of a mathematical model of the cosmos, we shall treat the galaxies as point masses or molecules forming a galactic gas and further assume that this gas behaves like the perfect fluid studied in section 22. In particular, its energy–momentum tensor will be supposed given by equation (22.21). At the present epoch, this gas is exceedingly rarefied and, since the random motions of the galaxies relative to the background black-body radiation (which provides a natural frame of reference) are of relatively small magnitude, the pressure associated with the gas is very low; thus, at this and later epochs, it will be permissible to neglect the pressure and to treat the gas as an incoherent dust cloud. However, during the early phases of cosmic evolution, the temperature and pressure are believed to have been very high indeed and the pressure terms cannot then be neglected; further, during these phases the contribution of the background radiation becomes significant and terms representing this contribution must also be included in the energy–momentum tensor.

We next assume that there are no privileged galactic observers, i.e. all observers moving with the galactic gas will be assumed to see the same large-scale process of evolution of the cosmos. This is the *cosmological principle*. The steady-state theory is based upon an extension called the *perfect cosmological principle*; this asserts that all galactic observers see the same large-scale state of the cosmos *at all times*. Observation supports the first principle but not the second.

If the cosmological principle is accepted and the perfect principle rejected, it is possible to define an absolute *cosmical time*, i.e. a way of assigning times to cosmic events which is independent of the observer. For all galactic observers will experience the same process of cosmic evolution and the various characteristic stages of this process can be allocated times according to some agreed scale. It is not necessary at this point in the argument to tie the scale to time measured by standard clocks; we only require that the later stages of an observer's experience be allocated times which are greater than the times allocated to earlier stages. Then, the cosmical time of any event can be defined unambiguously as the time recorded for the event by an adjacent galactic observer using the agreed time scale. Thus, the state of the cosmos at any epoch is now defined to be the set of events whose cosmical times are all equal to the cosmical time of the epoch. Since, at a given epoch, all galactic observers will be experiencing similar processes, the large-scale state of the cosmos at a given epoch must be homogeneous and isotropic for each such observer.

60. Spaces of constant curvature

At a given epoch, as we have just seen, the state of our cosmological model must be homogeneous and isotropic. In particular, the three-dimensional space in which the model is constructed must have these properties. The surface of a sphere is a two-dimensional space of this type embedded in $\mathscr{E}_3$ and it is obvious that a three-dimensional hypersphere embedded in $\mathscr{E}_4$ will have all the characteristics we need for our purpose. Also, just as the ordinary sphere includes the $\mathscr{E}_2$ plane as a special case when its radius becomes infinite, a hypersphere of infinite radius will correspond to $\mathscr{E}_3$, which is an especially simple case of a homogeneous and isotropic space. In addition, we shall be led quite naturally to consider a third class of such spaces which, like $\mathscr{E}_3$, but unlike the hypersphere, have infinite volume. These three types of space all have constant curvature scalars R and are therefore called *spaces of constant curvature*. All these spaces have positive-definite metrics, as they must have if their geometry is to be Euclidean over sufficiently small regions. It may be proved that there are no other Riemannian spaces having such metrics which are homogeneous and isotropic.

Let (x, y, z, u) be rectangular Cartesian coordinates in $\mathcal{E}_4$. Then, a hypersphere of radius S has equation

$$x^2 + y^2 + z^2 + u^2 = S^2 \tag{60.1}$$

Since u is determined in terms of x, y, z by this equation, a coordinate frame for points on this hypersurface can be constructed by first allocating coordinates (x, y, z) to the point having coordinates (x, y, z, u) in the Cartesian frame. Provided x, y, z are small by comparison with S, they will behave approximately like rectangular Cartesian coordinates in $\mathcal{E}_3$ and we shall therefore define quasi-spherical polar coordinates (r, θ, ϕ) by the usual transformation equations

$$x = r \sin \theta \cos \phi, \qquad y = r \sin \theta \sin \phi, \qquad z = r \cos \theta \tag{60.2}$$

Equations (60.1) and (60.2) give

$$u^2 = S^2 - r^2 \tag{60.3}$$

from which we find by differentiation that

$$du^2 = \frac{r^2}{u^2} \, dr^2 = \frac{r^2}{S^2 - r^2} \, dr^2 \tag{60.4}$$

Hence, the distance ds between the points (x, y, z, u) and $(x + dx, y + dy, z + dz, u + du)$ on the hypersphere is given by

$$
\begin{aligned}
ds^2 &= dx^2 + dy^2 + dz^2 + du^2 \\
&= dr^2 + r^2 \, (d\theta^2 + \sin^2 \theta d\phi^2) + r^2 \, dr^2/(S^2 - r^2) \\
&= \frac{S^2}{S^2 - r^2} \, dr^2 + r^2 \, (d\theta^2 + \sin^2 \theta d\phi^2)
\end{aligned} \tag{60.5}
$$

This is a metric for the hyperspherical $\mathcal{R}_3$. Clearly, by taking $1/S^2 = 0$, the metric reduces to that for $\mathcal{E}_3$ in ordinary spherical polars.

If the curvature scalar is calculated from this metric, it will be found that it equals $-6/S^2$, i.e. is constant. It is now evident that if S^2 is replaced by $-S^2$, the curvature scalar will still be constant with value $6/S^2$ and the space will remain homogeneous and isotropic. This is the third type of such a space; its metric is

$$ds^2 = \frac{S^2}{S^2 + r^2} \, dr^2 + r^2 \, (d\theta^2 + \sin^2 \theta d\phi^2) \tag{60.6}$$

Since these spaces are homogeneous and isotropic, the pole $r = 0$ can be taken to be any point and the axes from which θ and ϕ are measured can be taken in any pair of perpendicular directions. If r is small compared with S, both metrics (60.5) and (60.6) approximate to the spherical polar metric for $\mathcal{E}_3$, implying that the spaces are Euclidean over small regions.

Consider the circle $r = $ constant, $\theta = \frac{1}{2}\pi$, in the space with metric (60.5). The distance between neighbouring points ϕ, $\phi + d\phi$ on the circle is given by the metric to be $ds = rd\phi$ and the circumference of the circle is accordingly $2\pi r$. Along a radius of this circle $d\theta = d\phi = 0$ and the distance between points $r, r + dr$ is given to be

$$ds = Sdr/\sqrt{(S^2 - r^2)} \tag{60.7}$$

Integrating from $r = 0$ to r, the length of the radius is found to be

$$\rho = S \sin^{-1}(r/S) \tag{60.8}$$

Thus, $r = S \sin(\rho/S)$ and the circumference c is given in terms of the radius ρ by

$$c = 2\pi S \sin(\rho/S) \tag{60.9}$$

Since $\sin(\rho/S) < \rho/S$, c is smaller than $2\pi\rho$, which is the Euclidean result.

The formula (60.9) receives a simple interpretation in the allied case of the $\mathscr{R}_2$ which is the surface of an ordinary sphere of radius S. The quantities S, r, ρ are indicated in Fig. 8, from which the relationships just found are readily seen to be valid. Clearly, as ρ increases, r first increases until it achieves a maximum value S, and thereafter decreases until $\rho = \pi S$, when r becomes zero. It now appears that our coordinate system is ambiguous, in that two different points can have the same coordinates; this deficiency can be rectified by replacing r by ρ using the transformation equation (60.8), giving a new metric

$$ds^2 = d\rho^2 + S^2 \sin^2(\rho/S)(d\theta^2 + \sin^2\theta d\phi^2) \tag{60.10}$$

This transformation has also eliminated the singularity in the metric (60.5) at $r = S$. Like the Schwarzschild singularity, this is a property of the coordinate frame and evidently does not correspond to a singularity in the space itself.

Putting $\rho/S = \psi$, the metric (60.10) can also be expressed in the convenient form

$$ds^2 = S^2\{d\psi^2 + \sin^2\psi(d\theta^2 + \sin^2\theta d\phi^2)\} \tag{60.11}$$

(See Exercises 5, No. 33.)

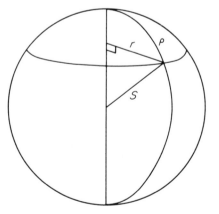

FIG. 8

In the case of the space with metric (60.6), r can assume all positive values and is not ambiguous. By putting $r = S \sinh \psi$, this metric can be transformed to

$$ds^2 = S^2 \{ d\psi^2 + \sinh^2 \psi (d\theta^2 + \sin^2 \theta d\phi^2) \} \qquad (60.12)$$

Another form for these metrics which will be specially important later is obtained by putting $r = S\sigma$. The metrics (60.5) and (60.6) can then both be expressed by

$$ds^2 = S^2 \left[\frac{d\sigma^2}{1 - k\sigma^2} + \sigma^2 (d\theta^2 + \sin^2 \theta d\phi^2) \right] \qquad (60.13)$$

where $k = 1$ in the case (60.5) and $k = -1$ in the case (60.6). The special Euclidean case can also be accommodated by permitting k to be zero. The new coordinate σ is dimensionless and, if $k = 1$, is restricted to values satisfying $0 \leqslant \sigma \leqslant 1$. For the other values of k, σ takes all positive values.

To calculate the volume of some region of an $\mathcal{R}_3$, let y^α ($\alpha = 1, 2, 3$) be geodesic rectangular Cartesian coordinates in the neighbourhood of some point P (section 39). Assuming that the metric is positive definite (as in the present case), the y^α will all be real. Then, if x^α are coordinates with respect to any other frame, the metric tensor in the x-frame will be given by

$$g_{\alpha\beta} = \frac{\partial y^\gamma}{\partial x^\alpha} \frac{\partial y^\delta}{\partial x^\beta} \delta_{\gamma\delta} = \frac{\partial y^\gamma}{\partial x^\alpha} \frac{\partial y^\gamma}{\partial x^\beta} \qquad (60.14)$$

Hence, if the Jacobian determinant $\partial(y^1, y^2, y^3)/\partial(x^1, x^2, x^3)$ is squared by multiplying its rows by its columns, it follows that

$$\left[\frac{\partial(y^1, y^2, y^3)}{\partial(x^1, x^2, x^3)} \right]^2 = |g_{\alpha\beta}| = g \qquad (60.15)$$

But, the volume δV of a small region A in the neighbourhood of P is given by

$$\delta V = \iiint_A dy^1 dy^2 dy^3 = \iiint_A \frac{\partial(y^1, y^2, y^3)}{\partial(x^1, x^2, x^3)} dx^1 dx^2 dx^3 \qquad (60.16)$$

Thus, the volume enclosed by the coordinate surfaces x^1, x^2, x^3, $x^1 + dx^1$, $x^2 + dx^2$, $x^3 + dx^3$ is

$$dV = \sqrt{g} \, dx^1 dx^2 dx^3 \qquad (60.17)$$

The formula for the volume V of a finite region F of $\mathcal{R}_3$ now follows, viz.

$$V = \iiint_F \sqrt{g} \, dx^1 dx^2 dx^3 \qquad (60.18)$$

In the case of the space with metric (60.5), $g = S^2 r^4 \sin^2 \theta / (S^2 - r^2)$ and the whole space has volume

$$2 \int_0^R dr \int_0^\pi d\theta \int_0^{2\pi} \frac{Sr^2 \sin \theta}{\sqrt{(S^2 - r^2)}} d\phi \qquad (60.19)$$

where the factor 2 is needed since the range $0 < r < S$ covers only half the sphere (see above). Performing the integrations, this volume is found to be $2\pi^2 S^3$.

It will be found that the total volume of the space with metric (60.6) is infinite, as in the Euclidean case.

61. The Robertson–Walker metric

In this section, we shall calculate a space–time metric for the cosmos in a frame formed by clocks moving with the galactic gas, all reading cosmical time x^4. As explained in section 45, the spatial coordinates x^α of each clock never change and the frame is therefore said to be co-moving with the gas.

Viewed from any one of the clocks, the cosmos is isotropic, i.e. it is impossible to specify any direction having special properties. Let g_{ij} be the metric tensor in this x-frame. Suppose we carry out a spatial coordinate transformation by relabelling the clocks, leaving their readings unchanged; such a transformation will take the form

$$\bar{x}^\alpha = f^\alpha(x^1, x^2, x^3), \quad \bar{x}^4 = x^4 \tag{61.1}$$

The transformed metric tensor is $\bar{g}_{ij}$ and we shall have

$$\bar{g}_{\alpha 4} = \frac{\partial x^i}{\partial \bar{x}^\alpha} \frac{\partial x^j}{\partial \bar{x}^4} g_{ij} = \frac{\partial x^\beta}{\partial \bar{x}^\alpha} g_{\beta 4} \tag{61.2}$$

This equation shows that $g_{\alpha 4}$ behaves as a covariant 3-vector with respect to spatial coordinate transformations and hence determines a special direction at every point of 3-space. This is contrary to our assumption of isotropy for galactic observers and we conclude that $g_{\alpha 4} = 0$ throughout space–time. Thus,

$$ds^2 = g_{\alpha\beta} dx^\alpha dx^\beta + g_{44} (dx^4)^2 \tag{61.3}$$

Now consider the events of a coordinate clock indicating the times x^4, $x^4 + dx^4$. Let $d\tau$ be the proper time interval between these events. Since the spatial coordinates of the clock never change, the metric (61.3) shows that

$$-c^2 d\tau^2 = g_{44} (dx^4)^2 \tag{61.4}$$

This equation determines the relationship between the cosmical time x^4 and the standard time τ shown on an atomic clock moving with the galactic observer. But this relationship must be independent of the galactic observer, since all are equivalent, and it follows that g_{44} can only depend on x^4. We can accordingly transform from x^4 to a new cosmical standard time t by a transformation

$$ct = \int \sqrt{(-g_{44})} \, dx^4 \tag{61.5}$$

so that the metric (61.3) reduces to

$$ds^2 = g_{\alpha\beta} dx^\alpha dx^\beta - c^2 dt^2 \tag{61.6}$$

Taking a section $t = $ constant of space–time at a particular cosmical time, an

$\mathscr{R}_3$ with metric

$$ds^2 = g_{\alpha\beta}dx^\alpha dx^\beta \tag{61.7}$$

is obtained. This is our model for the cosmos at this cosmical instant. By the cosmological principle, any galactic observer will find this $\mathscr{R}_3$ to be homogeneous and isotropic. Choosing himself as pole, he will therefore be able to define coordinates (σ, θ, ϕ) for which the metric (61.7) takes the form (60.13). Using this frame and the new cosmical time t, the metric (61.6) finally assumes the Robertson–Walker form

$$ds^2 = S^2\left[\frac{d\sigma^2}{1 - k\sigma^2} + \sigma^2(d\theta^2 + \sin^2\theta d\phi^2)\right] - c^2 dt^2 \tag{61.8}$$

S is a constant for the cosmos at any given time t, but will in general vary with t; it will be referred to as the *cosmic scale factor*; only when $k = 1$ can cosmic space be pictured as a hypersphere of radius S in $\mathscr{E}_4$.

It remains to check that the frame of reference is co-moving, as assumed at the outset. The galaxies will be falling freely in the gravitational field associated with the metric (61.8) and their world-lines must therefore be geodesics. Since we are supposing the spatial coordinates of a galaxy remain constant, $x^\alpha = $ constant along a galactic world-line. Substituting in the geodesic equations (43.5), the condition they are satisfied is found to be $\Gamma^\alpha_{44} = 0$. For the metric (61.8), this condition reduces to the requirement $\partial g_{44}/\partial x^\alpha = 0$, which is clearly true.

62. Hubble's constant and the deceleration parameter

The behaviour of the cosmic model derived from the Robertson–Walker metric will be determined when the value of k and the dependence of the cosmic scale factor S on the cosmical time t are known. From the physical data available today (1981), neither of these pieces of information can be derived with any degree of accuracy. It seems likely (see section 65) that $k = +1$ and that the universe is closed, i.e. of finite volume. In regard to $S(t)$, the value of its first derivative is known roughly, but even the sign of its second derivative is in doubt, although the general consensus of opinion is that it is negative, i.e. the cosmic expansion is slowing down. Instead of quoting values of these two derivatives, it is more convenient to work with the parameters

$$H = \dot{S}/S \tag{62.1}$$
$$q = -S\ddot{S}/\dot{S}^2 \tag{62.2}$$

H is called *Hubble's constant* and has reciprocal time dimension; q is called the *deceleration parameter* and is dimensionless. In the present epoch, the value of $1/H$ is often quoted to be about 1.8×10^{10} years, whereas the value of q is thought by some cosmologists to be about unity, although others would not exclude negative values.

At a fixed cosmical time t, the Robertson–Walker metric requires that ordinary space should have the metric (60.13). Any galaxy can be thought of as being placed at the pole $\sigma = 0$ and then the radial (or proper) distance of any other galaxy (σ, θ, ϕ) is given by

$$d = S \int_0^\sigma \frac{d\sigma}{\sqrt{(1 - k\sigma^2)}} = \alpha S \tag{62.3}$$

where $\alpha = \sin^{-1}\sigma$ if $k = 1$, $\alpha = \sigma$ if $k = 0$, and $\alpha = \sinh^{-1}\sigma$ if $k = -1$. Thus, the rate of recession of this galaxy from the galaxy at the origin is given by

$$\dot{d} = \alpha \dot{S} = Hd \tag{62.4}$$

This is *Hubble's law* that, at a given cosmical time, the rate of recession of any galaxy is proportional to its distance. H has the same value for all galactic observers at the time t, but will, in general, itself vary with t. Clearly, this law cannot be verified directly, since neither the distance d nor its rate of change are directly observable; in the next two sections, we shall derive an alternative relationship between associated quantities which can be measured.

63. Red shift of galaxies

Suppose that a galaxy $G(\sigma_1, \theta_1, \phi_1)$ is being observed through a telescope from the pole O. Successive crests of a light wave emitted by G at times $t_1, t_1 + dt_1$ are received at O at times $t_0, t_0 + dt_0$ respectively. The world-line of each crest is a radial null geodesic along which θ and ϕ remain constant (the reader should check that the equations of a null geodesic can be satisfied with θ and ϕ constant). The Robertson–Walker metric shows that along such a world-line,

$$\frac{d\sigma}{\sqrt{(1 - k\sigma^2)}} = -\frac{c}{S}dt \tag{63.1}$$

Integration of this equation for the motion of each crest yields the equations

$$\int_0^{\sigma_1} \frac{d\sigma}{\sqrt{(1 - k\sigma^2)}} = c \int_{t_1}^{t_0} \frac{dt}{S} = c \int_{t_1 + dt_1}^{t_0 + dt_0} \frac{dt}{S} \tag{63.2}$$

Since dt_0, dt_1 will be small, the last equation implies that

$$\frac{dt_0}{S(t_0)} = \frac{dt_1}{S(t_1)} \tag{63.3}$$

dt_1 is the period of the emitted wave as measured by a standard clock at G, and dt_0 is the period of the received wave as measured by a similar clock at O. Since wavelength is proportional to period, if λ_0, λ_1 are the wavelengths of the received and emitted light respectively,

$$\lambda_0 / \lambda_1 = S_0 / S_1 \tag{63.4}$$

where $S_0 = S(t_0)$ and $S_1 = S(t_1)$.

Thus, if the received light is redder than the emitted light, $\lambda_0 = \lambda_1 + \Delta\lambda_1$, where $\Delta\lambda_1$ is positive. The red-shift factor z is defined by the equation

$$z = \frac{\Delta\lambda_1}{\lambda_1} = \frac{\lambda_0}{\lambda_1} - 1 = \frac{S_0}{S_1} - 1 \qquad (63.5)$$

Clearly z is positive if $S_0 > S_1$, i.e. the universe is expanding.

If the observed galaxy is not too distant, $t_0 - t_1$ will be relatively small and we can expand $S(t_1)$ in a Taylor expansion thus:

$$S(t_1) = S(t_0) - (t_0 - t_1)\dot{S}(t_0) + \tfrac{1}{2}(t_0 - t_1)^2 \ddot{S}(t_0) + \dots$$
$$= S(t_0)\{1 - H_0(t_0 - t_1) - \tfrac{1}{2}q_0 H_0^2(t_0 - t_1)^2 + \dots\} \qquad (63.6)$$

where H_0, q_0 are the values of Hubble's constant and the deceleration parameter at the instant t_0 of observation. Substituting from the last equation in (63.5), we derive the result

$$z = H_0(t_0 - t_1) + (\tfrac{1}{2}q_0 + 1)H_0^2(t_0 - t_1)^2 + \dots \qquad (63.7)$$

By observing z for a number of galaxies and calculating $(t_0 - t_1)$ for each, this expansion provides a means of estimating the values of H and q at the present epoch. The calculation of $(t_0 - t_1)$ is considered in the next section.

64. Luminosity distance

If all galaxies possessed the same intrinsic luminosity, i.e. emitted light energy at the same rate, and if this luminosity were independent of the time, the observed or apparent luminosities of galaxies would depend upon their distances according to a calculable formula and an observation of the apparent luminosity would then provide us with a measure of the distance. Although there is considerable variation in the strengths of the galaxies as light sources, by confining observations to galaxies of a particular type and stage of evolution, this variation can be reduced and thus estimates of their distances can be obtained. Since galaxies tend to occur in clusters, once the distance of one member of a cluster has been found, the intrinsic luminosity of the other members can be determined, thus providing further useful information in regard to the probable intrinsic brightness of galaxies of other types; this information can then be utilized as a basis for later distance determinations.

Suppose we take the pole O of coordinates (σ, θ, ϕ) at some distant galaxy G whose luminosity is to be observed and let our point of observation A have coordinates $(\sigma, \tfrac{1}{2}\pi, 0)$. Photons emitted by G will have null geodesics as world-lines, along which θ and ϕ will be constant. Consider a photon which travels along the ray $\theta = \tfrac{1}{2}\pi$, $\phi = \varepsilon$, where ε is very small. This photon will ultimately arrive at the point $(\sigma, \tfrac{1}{2}\pi, \varepsilon)$ of closest approach to A when its distance from A will be given by equation (60.13) to be

$$ds = S_0 \sigma \varepsilon \qquad (64.1)$$

where S_0 is calculated at the time of arrival t_0 of the photon in the vicinity of A. It now follows that, if the telescope at A has aperture of radius a, the photon will be collected by the telescope provided $\varepsilon < a/S_0\sigma$. Thus, the telescope will collect all photons which left G along paths enclosed within a right circular cone of semi-vertical angle $a/S_0\sigma$. This cone embraces a solid angle

$$2\pi(1 - \cos\varepsilon) = \pi\varepsilon^2 = \frac{\pi a^2}{S_0^2\sigma^2} = \frac{\alpha}{S_0^2\sigma^2}, \tag{64.2}$$

where $\alpha = \pi a^2$ is the telescope's aperture; it follows that the proportion of photons leaving G which are collected at A is $\alpha/(4\pi S_0^2\sigma^2)$.

According to equation (63.4), if v_1 is the frequency of a photon when it leaves G at time t_1, its frequency v_0 on arrival at A at time t_0 is given by $v_0 = S_1 v_1/S_0$. But the energy of a photon is related to its frequency by the formula $E = hv$, where h is Planck's constant. Thus, the energy of the photon is also reduced by a factor S_1/S_0. Further, equation (63.3) indicates that the photons emitted from G over a time interval dt_1 arrive at A over the longer time interval $dt_0 = S_0\,dt_1/S_1$. Hence, the rate of reception of light energy is additionally reduced by a factor S_1/S_0. If, therefore, L is the *intrinsic luminosity* of G, i.e. the total rate at which it emits radiation, the rate at which light energy is collected by the telescope at A is given by

$$L \times \frac{\alpha}{4\pi S_0^2\sigma^2} \times \frac{S_1}{S_0} \times \frac{S_1}{S_0} = \frac{L\alpha S_1^2}{4\pi\sigma^2 S_0^4}. \tag{64.3}$$

The *apparent luminosity* l of a celestial object is defined to be the rate at which light energy from the object flows across unit area normal to the line of sight at the point of observation. The last result shows that for G,

$$l = LS_1^2/(4\pi\sigma^2 S_0^4) \tag{64.4}$$

If space were Euclidean and G were stationary at a distance d from A, we should have

$$l = L/(4\pi d^2) \tag{64.5}$$

Substituting the actual value of l from (64.4) and solving for d, we find

$$d = \sigma S_0^2/S_1 \tag{64.6}$$

This result is termed the *luminosity distance* of G.

Returning to equation (64.4), we shall obtain an expansion for l in powers of $(t_0 - t_1)$. Equation (63.2) shows that

$$\int_0^\sigma \frac{d\sigma}{\sqrt{(1 - k\sigma^2)}} = c\int_{t_1}^{t_0} \frac{dt}{S} \tag{64.7}$$

Assuming σ is relatively small (i.e. G is not too distant), the integrand of the left-hand member of this equation can be expanded in powers of σ to give

$$\int_0^\sigma \frac{d\sigma}{\sqrt{(1 - k\sigma^2)}} = \sigma + \frac{1}{6}k\sigma^3 + \ldots \tag{64.8}$$

Also, equation (63.6) leads to the result

$$\frac{1}{S(t)} = \frac{1}{S(t_0)} \{1 + H_0(t_0 - t) + \ldots\} \tag{64.9}$$

Hence, the right-hand member of equation (64.7) can be expanded in the form

$$c \int_{t_1}^{t_0} \frac{dt}{S} = \frac{c}{S_0} \{(t_0 - t_1) + \tfrac{1}{2}H_0(t_0 - t_1)^2 + \ldots\} \tag{64.10}$$

Equating the expansions (64.8) and (64.10), we see that, to the first order of small quantities, $\sigma = c(t_0 - t_1)/S_0$. It follows that, to the second order in $(t_0 - t_1)$,

$$\sigma = \frac{c}{S_0} \{(t_0 - t_1) + \tfrac{1}{2}H_0(t_0 - t_1)^2 + \ldots\} \tag{64.11}$$

Substitution from equations (63.6) and (64.11) in equation (64.4) now yields the expansion

$$l = \frac{L}{4\pi c^2 (t_0 - t_1)^2} \{1 - 3H_0(t_0 - t_1) + \ldots\} \tag{64.12}$$

Finally, equation (63.7) shows that to the first order $(t_0 - t_1) = z/H_0$. It follows from the same equation, therefore, that

$$H_0(t_0 - t_1) = z - (\tfrac{1}{2}q_0 + 1)z^2 + \ldots \tag{64.13}$$

Substituting for $(t_0 - t_1)$ from this equation into equation (64.12) accordingly leads to the expansion

$$l = \frac{LH_0^2}{4\pi c^2 z^2} \{1 + (q_0 - 1)z + \ldots\} \tag{64.14}$$

The importance of this last equation is that it relates two observable quantities l and z. By fitting it as closely as possible to the available data, estimates of H_0 and q_0 have been obtained. A more precise relationship follows from dynamical considerations (see Exercises 7, No. 5).

65. Cosmic dynamics

To gain further information in regard to the probable values of the functions $H(t)$ and $q(t)$ at cosmical times in the remote past and future, it is necessary to determine the equations of motion of the cosmos. These will be provided by Einstein's equations of gravitation. It is therefore necessary first to calculate the energy–momentum tensor for the galactic gas.

The 4-velocity of a particle whose world-line in the x-frame has equations $x^i = x^i(\tau)$ ($\tau = s/ic$ being proper time measured by a standard clock moving with the particle) has been defined (section 50) to be the contravariant vector V^i given by

$$V^i = \frac{dx^i}{d\tau} \tag{65.1}$$

If the frame is inertial and Minkowski coordinates are used, this definition is in agreement with that adopted in the special theory (section 15).

Now consider a perfect fluid. At any point in the fluid, we can construct a freely falling frame which is locally inertial and in which Minkowski coordinates can be defined. The special theory is valid in any such frame and the fluid's proper density of proper mass μ_{00} and pressure p can be defined as 4-invariants (section 22). The energy–momentum tensor in the frame then follows from equation (22.21). In the general x-frame, this tensor must therefore be given by the equation

$$T^{ij} = (\mu_{00} + p/c^2) V^i V^j + p g^{ij} \tag{65.2}$$

for this is a tensor equation and reduces to the valid equation (22.21) in the freely falling Minkowski frame.

Along a world-line of a particle of the galactic gas, the coordinates (σ, θ, ϕ) remain constant and hence, from the metric equation (61.8), we deduce that $\tau = t$. Thus, the 4-velocity of this particle has components $(V^i) = (0, 0, 0, 1)$ in the Robertson–Walker frame. The non-zero contravariant components of the metric tensor are

$$g^{11} = (1 - k\sigma^2)/S^2, \quad g^{22} = 1/S^2\sigma^2, \quad g^{33} = \operatorname{cosec}^2\theta/S^2\sigma^2, \quad g^{44} = -1/c^2 \tag{65.3}$$

We can now calculate the non-zero components of the energy–momentum tensor for the galactic gas from equation (65.2); they are

$$\left.\begin{aligned}
T^{11} &= p(1 - k\sigma^2)/S^2 \\
T^{22} &= p/S^2\sigma^2 \\
T^{33} &= p \operatorname{cosec}^2\theta/S^2\sigma^2 \\
T^{44} &= \mu
\end{aligned}\right\} \tag{65.4}$$

where we have deleted the subscripts in μ_{00} for convenience. The covariant components now follow immediately, viz.

$$\left.\begin{aligned}
T_{11} &= pS^2/(1 - k\sigma^2) \\
T_{22} &= pS^2\sigma^2 \\
T_{33} &= pS^2\sigma^2 \sin^2\theta \\
T_{44} &= \mu c^4
\end{aligned}\right\} \tag{65.5}$$

The non-zero components of the Ricci tensor for the Robertson–Walker metric may be verified to be

$$R_{11} = -P/(1 - k\sigma^2), \quad R_{22} = -P\sigma^2, \quad R_{33} = -P\sigma^2 \sin^2\theta, \quad R_{44} = 3\ddot{S}/S \tag{65.6}$$

where

$$P = 2k + (S\ddot{S} + 2\dot{S}^2)/c^2 \tag{65.7}$$

dots denoting differentiations with respect to t.

The reader may also verify that the curvature scalar is given by

$$R = R_i^i = -6(S\ddot{S} + \dot{S}^2 + kc^2)/c^2 S^2 \qquad (65.8)$$

We can now construct Einstein's equations (47.16) (covariant form); only two distinct equations emerge, viz.

$$2S\ddot{S} + \dot{S}^2 + kc^2 - c^2 \Lambda S^2 = -\kappa c^2 p S^2 \qquad (65.9)$$
$$3(\dot{S}^2 + kc^2) - c^2 \Lambda S^2 = \kappa c^4 \mu S^2 \qquad (65.10)$$

Together with an equation of state $p = p(\mu)$, these equations are sufficient to determine the unknown functions S, p and μ.

Before attempting to integrate these equations, certain important conclusions can be reached by a direct study.

Equation (65.10) can be written in the form

$$\frac{3c^2}{S^2} k = \kappa c^4 \mu - (3H^2 - c^2 \Lambda) \qquad (65.11)$$

from which it follows that $k = +1$ if

$$\mu > \mu_c = (3H^2 - c^2 \Lambda)/\kappa c^4 \qquad (65.12)$$

and $k = -1$ if the reverse inequality is true. μ_c is therefore a critical density (the *closure density*) for the galactic gas, determining whether the universe is of finite or infinite volume. Assuming $\Lambda = 0$ and taking $H = 1.8 \times 10^{-18}$ (at the present epoch), $\kappa c^4 = 8\pi G = 1.67 \times 10^{-9}$ (all SI units), we calculate that

$$\mu_c = 6 \times 10^{-27} \, \text{kg m}^{-3} \qquad (65.13)$$

If the matter in the galaxies were spread uniformly over the whole of space at the present epoch, it is estimated that its density would be $\mu = 3 \times 10^{-28} \, \text{kg m}^{-3}$. Thus $\mu < \mu_c$ and this datum suggests that the universe is open. However, the presence of a very tenuous intergalactic dust or gas, far below the limit of possible observation, could easily reverse this conclusion. We accordingly seek further information from equation (65.9).

This equation can be written

$$\frac{c^2}{S^2} k = (2q - 1) H^2 + c^2 \Lambda - \kappa c^2 p \qquad (65.14)$$

Again taking $\Lambda = 0$ and assuming p to be negligible, we conclude that $k = +1$ if $q > \frac{1}{2}$. Unfortunately, the red shift–luminosity relationship, from which the value of q is derivable (see section 64), has not been established with sufficient certainty to decide for or against this inequality. However, the data available tends to support it.

66. Model universes of Einstein and de Sitter

The first solution of the dynamical equations (65.9) and (65.10) was suggested by Einstein himself. His proposal was made some years before Hubble published his observations relating to the recession of the galaxies, and the possibility that the universe was in a state of expansion was not considered by astronomers at that time. Einstein's universe was therefore static and the equations were satisfied by taking S to be constant. Pressure and density had then to satisfy the equations

$$\kappa p = \Lambda - \frac{k}{S^2} \tag{66.1}$$

$$\kappa c^2 \mu = \frac{3k}{S^2} - \Lambda \tag{66.2}$$

Clearly, if $\Lambda = 0$, since p and μ cannot be negative, the only possible solution is for p, μ and k all to be zero, i.e. an empty, infinite, Euclidean cosmos. It was in order to be able to reject this solution that Einstein introduced the cosmical constant term into his equation of gravitation.

Adding equations (66.1) and (66.2), we find

$$\frac{2k}{S^2} = \kappa(p + c^2 \mu) \tag{66.3}$$

proving that k must be positive, i.e. $k = +1$ and the universe is closed. Equations (66.1) and (66.2) now show that Λ must satisfy the inequalities

$$\frac{1}{S^2} \leqslant \Lambda \leqslant \frac{3}{S^2} \tag{66.4}$$

S is the radius of the universe, which is certainly very large, indicating that Λ will be quite inappreciable except for phenomena on the cosmic scale. Following upon Hubble's discovery, Einstein abandoned his model and with it, the cosmical constant. Since then, Λ has been put to zero in most cosmological investigations and we shall follow this practice in the remaining sections of this chapter (except in certain exercises at the end).

The necessity for including the cosmical constant term if the universe is to be static, follows from very elementary considerations. Without it, an initially static cosmos will collapse under the gravitational attractions of its constituent galaxies. If Λ is positive, the term implies the existence of a counterbalancing long-range repulsive force between the galaxies, which increases with their distance of separation (Exercises 6, No. 40). However, although these two opposing gravitational forces can result in equilibrium, putting $p = 0$ to match the present state of the cosmos and hence $\Lambda = 1/S^2$, it may be proved that the equilibrium is unstable (Exercises 7, No. 8).

Following upon Einstein's proposed model, de Sitter derived another which is of some historical interest. In this context, his model is most conveniently constructed by adopting as our basic assumption that Hubble's constant does not change with time, i.e.

$$\dot{S}/S = H = \text{constant} \tag{66.5}$$

Integrating, we get

$$S = A \exp Ht \tag{66.6}$$

where A is the value of S at some arbitrarily chosen origin of cosmical time. Substituting in equations (65.9) and (65.10), we find that

$$\kappa p = \Lambda - \frac{3H^2}{c^2} - \frac{k}{A^2} \exp(-2Ht) \tag{66.7}$$

$$\kappa c^2 \mu = -\Lambda + \frac{3H^2}{c^2} + \frac{3k}{A^2} \exp(-2Ht) \tag{66.8}$$

As $t \to \infty$,

$$\kappa p \to \Lambda - 3H^2/c^2, \qquad \kappa c^2 \mu \to -\Lambda + 3H^2/c^2 \tag{66.9}$$

Neither of these limits can be negative, so we must require that

$$\Lambda = 3H^2/c^2 \tag{66.10}$$

Thus, equations (66.7) and (66.8) lead to the result

$$3p + c^2\mu = 0 \tag{66.11}$$

from which we conclude that

$$p = 0, \ \mu = 0 \tag{66.12}$$

and, therefore, that $k = 0$.

Thus, de Sitter's universe is Euclidean, but being empty is only of academic interest. Although the model appears to be in a state of expansion, since no matter is present, this phenomenon has no physical basis; indeed, it is possible to derive a static metric for the model by carrying out certain tranformations on σ and t (see Exercises 7, No. 3). If the galaxies are represented by freely falling particles having negligible mass and constant spatial coordinates, an expanding infinite cosmos is obtained; however, each galactic observer is limited to a finite universe, since he cannot penetrate beyond a certain distance called his *event horizon* (see Exercises 7, No. 4).

67. Friedmann universes

These models are constructed from equations (65.9) and (65.10) by putting $p = 0$, i.e. by treating the galactic gas as an incoherent dust cloud. We shall also put Λ to zero.

We first note that equations (65.11) and (65.14) reduce to the forms

$$\kappa c^4 \mu = 3H^2 + 3kc^2/S^2 \tag{67.1}$$

$$(2q - 1)H^2 = kc^2/S^2 \tag{67.2}$$

It follows that

$$\kappa c^4 \mu = 6qH^2 \tag{67.3}$$

and hence that q can never be negative.

Equation (65.9) gives

$$2S\ddot{S} + \dot{S}^2 + kc^2 = 0 \tag{67.4}$$

Writing this equation in the form

$$\frac{d}{dt}(S\dot{S}^2) = -kc^2\dot{S} \tag{67.5}$$

we can integrate and obtain

$$S\dot{S}^2 = c^2(D - kS) \tag{67.6}$$

D being constant.

Substitution in equation (65.10) now yields

$$\mu S^3 = 3D/\kappa c^2 = \text{constant} \tag{67.7}$$

showing that the constant D must be positive. This is the equation of conservation of mass (see Exercises 7, No. 2). In the case of a finite cosmos, the total volume is known to be $2\pi^2 S^3$ and the total proper mass is accordingly $2\pi^2 S^3 \mu$, which is conserved. This is to be expected, since there is no interaction between the particles of the dust cloud and their proper masses therefore never change.

We shall next integrate equation (67.6), with k taking its three possible values, separately.

(i) If $k = 1$, the equation is

$$S\dot{S}^2 = c^2(D - S) \tag{67.8}$$

Changing the dependent variable from S to u by the transformation

$$S = \tfrac{1}{2}D(1 - \cos u) \tag{67.9}$$

we have $\dot{S} = \tfrac{1}{2}D\dot{u}\sin u$ and equation (67.8) leads to the equation

$$\tfrac{1}{2}D(1 - \cos u)\dot{u} = c \tag{67.10}$$

This integrates immediately to the form

$$ct = \tfrac{1}{2}D(u - \sin u) \tag{67.11}$$

choosing t to be zero when $u = 0$ (i.e. the origin of cosmical time is taken to be the instant when $S = 0$). Equations (67.9) and (67.11) determine S as a function of t;

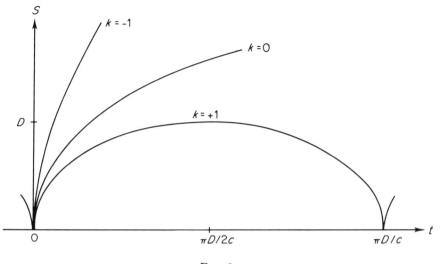

FIG. 9

since these equations are the well-known parametric equations for a cycloid, the plot of S against ct must take this form (see Fig. 9).

Our conclusion is that the finite universe will expand from a singularity at $t = 0$ to a maximum radius D when $u = \pi$ and $t = \pi D/2c$, and will then contract back to the singular state at $u = 2\pi$, $t = \pi D/c$. The actual behaviour in the vicinity of the singularity is, of course, not predicted by our analysis, since p has been neglected and the contribution of the radiation ignored.

(ii) If $k = 0$, equation (67.6) gives

$$S\dot{S}^2 = c^2 D \tag{67.12}$$

The variables separate and integration yields

$$S^3 = 9c^2 D t^2/4 \tag{67.13}$$

again taking $t = 0$ at $S = 0$. This universe expands from a singularity and continues to do so indefinitely (Fig. 9). Space is Euclidean and the cosmos is open.

(iii) If $k = -1$, equation (67.6) takes the form

$$S\dot{S}^2 = c^2 (D + S) \tag{67.14}$$

As in the case $k = 1$, we change the variable to u, this time by the transformation

$$S = \tfrac{1}{2} D(\cosh u - 1) \tag{67.15}$$

and integrate to find that

$$ct = \tfrac{1}{2} D(\sinh u - u) \tag{67.16}$$

taking u and t to be zero at $S = 0$. As in the previous case, the initial expansion is never reversed and the universe is open (Fig. 9).

It remains to calculate two constants of integration, viz. D and the value t_0 of t in the present epoch, in terms of H_0 and q_0.

Putting present values into equation (67.2), if $k \neq 0$ we get

$$S_0^2 = \frac{kc^2}{(2q_0 - 1)H_0^2} \tag{67.17}$$

Equation (67.6) can be written in the form

$$H^2 = c^2(D - kS)/S^3 \tag{67.18}$$

Thus,

$$D = \frac{1}{c^2} S_0^3 H_0^2 + kS_0 \tag{67.19}$$

If $k \neq 0$, equation (67.17) now shows that

$$\left. \begin{aligned} D &= \frac{2c}{H_0} q_0 (2q_0 - 1)^{-3/2}, \; q_0 > \tfrac{1}{2}, \, k = 1 \\ &= \frac{2c}{H_0} q_0 (1 - 2q_0)^{-3/2}, 0 < q_0 < \tfrac{1}{2}, \, k = -1. \end{aligned} \right\} \tag{67.20}$$

If $k = 0$, equation (67.2) requires that $q = \tfrac{1}{2}$ and equation (67.19) gives

$$D = S_0^3 H_0^2 / c^2 \tag{67.21}$$

S_0, H_0 are independent parameters for this model. However, S_0 is superfluous and may be eliminated by transformation of the coordinate σ to $\sigma' = S_0 \sigma$; S_0 then disappears from the metric. Equivalently, we may take $S_0 = 1$ and write equation (67.13) in the final form

$$4S^3 = 9H_0^2 t^2 \tag{67.22}$$

The present age t_0 of the universe is now calculable for the three types of model.

(i) If $k = 1$, $q_0 > \tfrac{1}{2}$, equation (67.9) shows that

$$\cos u_0 = 1 - 2S_0/D = \frac{1}{q_0} - 1 \tag{67.23}$$

having used equations (67.17) and (67.20). Equation (67.11) now yields the result

$$t_0 = \frac{1}{H_0} [q_0 (2q_0 - 1)^{-3/2} \cos^{-1}(q_0^{-1} - 1) - (2q_0 - 1)^{-1}] \tag{67.24}$$

the inverse cosine being taken in the first or second quadrant. If values $q_0 = 1$, $1/H_0 = 1.8 \times 10^{10}$ years are substituted, we find $t_0 = 10^{10}$ years very nearly, i.e. ten billion years.

(ii) If $k = -1$, $0 < q_0 < \tfrac{1}{2}$, a similar calculation leads to the result

$$t_0 = \frac{1}{H_0} [(1 - 2q_0)^{-1} - q_0 (1 - 2q_0)^{-3/2} \cosh^{-1}(q_0^{-1} - 1)] \tag{67.25}$$

Suppose we calculate q_0 from equation (67.3), taking $\mu = 3 \times 10^{-28}$ kg m^{-1} (i.e. assuming all the matter has been attracted into the galaxies) and $H_0 = 1.8 \times 10^{-18}$ s^{-1}. Then $q_0 = 0.025$ and the formula (67.24) gives $t_0 = 1.6 \times 10^{10}$ years (16 billion years).

(iii) Finally, if $k = 0$, $q = \frac{1}{2}$, putting $S = 1$ into equation (67.22) gives

$$t_0 = 2/(3H_0) \tag{67.26}$$

With $1/H_0 = 1.8 \times 10^{10}$ years, this makes the present age of the cosmos to be 12 billion years.

For the early stage of cosmic expansion, the model's failure to include the contribution of the radiation renders it invalid. However, since this stage is expected to be short (about 10^6 years), the corrections which need to be made to the values of t_0 calculated above to allow for this failure are negligible.

The stage of radiation dominance is studied in the next section.

68. Radiation model

During the early stages of cosmic expansion, it is believed that the dominating factor was the electromagnetic radiation. In this section, we shall study a simplified cosmological model, containing radiation alone, which will provide a first approximation to this early state of the universe.

We shall assume that the radiation is isotropic for each galactic observer and has energy density U (the same for all observers at a cosmical time t). Such an observer is in a state of free fall and can be equipped with rectangular axes and associated standard clocks, which behave locally like an inertial frame. Minkowski coordinates y^i can be defined in this frame and, with respect to these, the observer's 4-velocity is $\mathbf{V} = (0, ic)$ (section 15).

In this y-frame, the results of the special theory are applicable and, in particular, equation (29.5) defines the energy–momentum tensor S_{ij} for the radiation. We shall assume that the time variations of the field components E_α, H_α of this radiation are quite random and that there is no correlation between these components. Thus, if $\alpha \neq \beta$,

$$m(E_\alpha E_\beta) = 0, \qquad m(H_\alpha H_\beta) = 0 \tag{68.1}$$

where $m(\square)$ denotes the mean value. Also, since the radiation is isotropic, $m(E_\alpha^2)$ and $m(H_\alpha^2)$ will be independent of α. Hence, taking mean values in equation (29.16), we find

$$\varepsilon_0 m(E_\alpha^2) + \mu_0 m(H_\alpha^2) = \frac{2}{3} U \tag{68.2}$$

U being the mean energy density of the radiation.

We can now calculate the mean values of the components of S_{ij}. If $\alpha \neq \beta$, it follows from equations (29.14) and (68.1) that $m(S_{\alpha\beta}) = 0$. Also, since the radiation is in a steady state, the rate of energy flow in any direction is zero and, hence, the mean values of the components of the Poynting vector all vanish; i.e. $m(S_{\alpha 4}) = 0$. It is clear, therefore, that only the means of the diagonal components $S_{11}, S_{22}, S_{33}, S_{44}$ are non-zero and, for these, equations (29.13) and (29.16) give

$$m(S_{11}) = m(S_{22}) = m(S_{33}) = \tfrac{1}{2}\varepsilon_0 m(E_\alpha^2) + \tfrac{1}{2}\mu_0 m(H_\alpha^2) = \frac{1}{3}U \qquad (68.3)$$

$$m(S_{44}) = -U \qquad (68.4)$$

A tensor equation for S^{ij}, valid in any frame, can now be written down, viz.

$$S^{ij} = \frac{4\,U}{3\,c^2}\,V^i V^j + \frac{1}{3}g^{ij}U \qquad (68.5)$$

where V^i is the 4-velocity of the local galactic observer. U is uniquely defined and is thus a 4-invariant. Hence, this equation certainly defines a contravariant tensor. The equation is easily verified to give the mean components just calculated in the y-frame. It is therefore valid in all frames. If the equation is compared with equation (65.2) for a perfect fluid, it will be seen that the radiation behaves like a perfect fluid of density U/c^2 and pressure $U/3$.

In the Robertson–Walker frame, $(V^i) = (0, 0, 0, 1)$. Hence, the non-zero components of the energy–momentum tensor in this frame are:

$$\left.\begin{aligned} S^{11} &= U(1 - k\sigma^2)/3S^2 \\ S^{22} &= U/(3S^2\sigma^2) \\ S^{33} &= U\,\mathrm{cosec}^2\,\theta/(S^2\sigma^2) \\[4pt] S^{44} &= U/c^2 \end{aligned}\right\} \qquad (68.6)$$

Using the components of the Ricci tensor already calculated in section 65, the Einstein equations for the model can now be calculated. They prove to be

$$2S\ddot{S} + \dot{S}^2 + kc^2 - c^2\Lambda S^2 = -\tfrac{1}{3}\kappa c^2 U S^2 \qquad (68.7)$$

$$3(\dot{S}^2 + kc^2) - c^2\Lambda S^2 = \kappa c^2 U S^2 \qquad (68.8)$$

Since S will be comparatively small during this phase, even if Λ is non-zero, the terms containing this constant will be negligible. We accordingly put $\Lambda = 0$. Also, if $k \neq 0$, the term kc^2 will be negligible by comparison with the terms $2S\ddot{S}$ and $\dot{S}^2$. For, during the matter-dominated phase, equation (67.6) shows that

$$\dot{S}^2 = c^2\left(\frac{D}{S} - k\right) \qquad (68.9)$$

and kc^2 is small by comparison with $\dot{S}^2$ provided $D/S \gg 1$. But

$$\frac{D}{S} = \frac{D}{S_0}\frac{S_0}{S} = \frac{2q_0}{|2q_0 - 1|} \cdot \frac{S_0}{S} \qquad (68.10)$$

having used equations (67.17) and (67.20). At the beginning of this phase, S_0/S is large and it follows that D/S is large. Thus, kc^2 is negligible by comparison with

$\dot{S}^2$ and, since $\dot{S}$ takes even larger values during the earlier radiation dominated phase, we shall neglect kc^2 during this phase. We have proved, therefore, that in this early stage of the cosmic expansion, the behaviour of the cosmos is independent of the values of Λ and k.

Eliminating U between equations (68.7) and (68.8), it is now found that

$$S\ddot{S} + \dot{S}^2 = \frac{d}{dt}(S\dot{S}) = 0 \tag{68.11}$$

Two integrations now yield the results

$$\dot{S} = A/S, \qquad S^2 = 2At \tag{68.12}$$

where A is constant. Substituting in equation (68.8), we find

$$\kappa c^2 U = 3\dot{S}^2/S^2 = 3/(4t^2) \tag{68.13}$$

If we assume that during this phase the radiation is in thermal equilibrium with the matter present, then it will possess a black-body spectrum and the temperature T will be given in terms of the energy density U by Stefan's law, viz.

$$U = aT^4 \tag{68.14}$$

where $a = 7.5 \times 10^{-16}\,\mathrm{J\,m^{-3}\,K^{-4}}$ is the Stefan–Boltzmann constant. Thus, equation (68.13) leads to the following formula for T:

$$T = \left(\frac{3c^2}{32\pi Ga}\right)^{1/4} t^{-1/2} = 1.52t^{-1/2} \times 10^{10}\,\mathrm{K} \tag{68.15}$$

One second after the inception of the expansion, this formula indicates a temperature of $1.52 \times 10^{10}\,\mathrm{K}$ for the cosmos.

69. Particle and event horizons

Equation (64.7) determines the coordinate σ of an object which is observed in our telescopes at the present epoch t_0, by light which was emitted by the object at time t_1. Since t_1 cannot be less than the time $t = 0$ at which the cosmic expansion commenced (accepting the big-bang hypothesis), the most distant object which we can observe today has coordinate σ given by

$$\int_0^\sigma \frac{d\sigma}{\sqrt{(1 - k\sigma^2)}} = c \int_0^{t_0} \frac{dt}{S} \tag{69.1}$$

The proper distance of such an object is therefore

$$d_H = S_0 \int_0^\sigma \frac{d\sigma}{\sqrt{(1 - k\sigma^2)}} = cS_0 \int_0^{t_0} \frac{dt}{S} \tag{69.2}$$

d_H is said to be the proper distance of the *particle horizon*.

In the case of the Friedmann model with $k = +1$, equations (67.9) and (67.11) give

$$c\frac{dt}{S} = \frac{\frac{1}{2}D(1-\cos u)du}{\frac{1}{2}D(1-\cos u)} = du \tag{69.3}$$

and, thus,

$$d_H = S_0 u_0 = \frac{c}{H_0}(2q_0-1)^{-1/2}\cos^{-1}\left(\frac{1}{q_0}-1\right) \tag{69.4}$$

by equations (67.17) and (67.23).

The reader is left to obtain the results

$$d_H = \frac{2c}{H_0}, \qquad k = 0 \tag{69.5}$$

$$d_H = \frac{c}{H_0}(1-2q_0)^{-1/2}\cosh^{-1}\left(\frac{1}{q_0}-1\right),$$
$$\text{if } k = -1 \tag{69.6}$$

for the other values of k, in a similar manner.

In particular, if $k = +1$, $q_0 = 1$, $1/H_0 = 1.8 \times 10^{10}$ years, we calculate that $d_H = 2.8 \times 10^{10}$ light years.

Equation (64.7) can be utilized in an alternative manner. If an event occurs at the point with coordinate σ at the present epoch t_0, we shall observe it at time t_1 provided

$$\int_0^\sigma \frac{d\sigma}{\sqrt{(1-k\sigma^2)}} = c\int_{t_0}^{t_1} \frac{dt}{S} \tag{69.7}$$

In the case of the Friedmann model with $k = +1$, we must have $t_1 < \pi D/c$, since the cosmos collapses to a point at time $\pi D/c$; for $k = 0$ or -1, t_1 can be arbitrarily large. Thus, the proper distance in the present epoch of the most distant event we can ever hope to see is given by

$$d_E = cS_0\int_{t_0}^{t_{max}} \frac{dt}{S} \tag{69.8}$$

where $t_{max} = \pi D/c$ if $k = +1$, and $t_{max} = \infty$ if $k = 0$ or -1. If $d_E = \infty$, then all events happening in the present epoch will ultimately be observed in our telescopes. d_E is termed the proper distance of the *event horizon*.

In the case $k = +1$, we calculate that

$$d_E = S_0(u_{max}-u_0) = S_0(2\pi - u_0)$$
$$= \frac{c}{H_0}(2q_0-1)^{-1/2}\cos^{-1}\left(\frac{1}{q_0}-1\right) \tag{69.9}$$

provided the inverse cosine is taken in the range $(\pi, 2\pi)$. With $q_0 = 1$, $1/H_0 = 1.8 \times 10^{10}$ years, it will be found that $d_E = 8.4 \times 10^{10}$ light years. Light from events occurring at a greater proper distance will not have reached us before the cosmos collapses to a singularity.

The cases $k = 0$ and -1 both give an infinite value of d_E and all events happening today will, in these models, ultimately be observable on the earth.

Exercises 7

1. Transforming to a new radial coordinate r by the equation $\sigma = r/(1 + \frac{1}{4}kr^2)$, obtain the Robertson–Walker metric in the form

$$ds^2 = \left(\frac{S}{1 + \frac{1}{4}kr^2}\right)^2 \{dr^2 + r^2(d\theta^2 + \sin^2\theta\, d\phi^2)\} - c^2 dt^2$$

2. From the conservation equation $T^{4i}{}_{;i} = 0$, obtain the equation

$$\frac{d}{dt}(\mu S^3) + \frac{3}{c^2} S^2 \dot{S} p = 0$$

Deduce that, if $p = 0$, then $\mu \propto 1/S^3$.

3. Show that the de Sitter metric

$$ds^2 = A^2 \exp(2HT)(d\sigma^2 + \sigma^2 d\theta^2 + \sigma^2 \sin^2\theta\, d\phi^2) - c^2 dt^2$$

can be transformed to the static form

$$ds^2 = \frac{dr^2}{1 - H^2 r^2/c^2} + r^2(d\theta^2 + \sin^2\theta\, d\phi^2) - c^2(1 - H^2 r^2/c^2)dT^2$$

by the transformation equations

$$\sigma = \frac{r \exp(-HT)}{A\sqrt{(1 - H^2 r^2/c^2)}}, \qquad t = T + \frac{1}{2H}\log(1 - H^2 r^2/c^2).$$

4. A photon is emitted from the point (σ, θ, ϕ) along the radius to the origin at time t in the de Sitter universe whose metric is given in the previous exercise. Show that the time taken to reach the origin is

$$-\frac{1}{H}\log\left[1 - \frac{HA}{c}\sigma \exp(Ht)\right].$$

Hence show that if the proper distance of the point from the origin at the time t is greater than c/H, the photon will never arrive at O.

5. Show that, for all the Friedmann models with $\Lambda = 0$, $p = 0$, the luminosity distance d_L of a galaxy whose red shift is z is given by

$$d_L = \frac{c}{H_0 q_0^2}[q_0 z + (q_0 - 1)\{\sqrt{(2q_0 z + 1)} - 1\}].$$

If z is small, verify the expansion (64.14).

6. Show that if Λ is not assumed to vanish in the Friedmann model, then $S(t)$ satisfies

$$S\dot{S}^2 = c^2(D - kS + \tfrac{1}{3}\Lambda S^3)$$

where D is a matter density parameter defined by the equation $\kappa c^2 \mu S^3 = 3D$. Show that the special case $k = 0$, $D = 0$ yields the de Sitter universe.

7. Sketch the graph of $S\dot{S}^2$ against S for the Friedmann model with cosmical constant (previous exercise) and deduce that (i) S increases from zero to a maximum value and then decreases back to zero in the cases: (a) $k = +1$, $\Lambda < 4/(9D^2)$, (b) $k = 0$, $\Lambda < 0$, (c) $k = -1$, $\Lambda < 0$; (ii) S increases steadily from zero and tends to infinity in the cases: (a) $k = +1$, $\Lambda > 4/(9D^2)$, (b) $k = 0$, $\Lambda \geqslant 0$, (c) $k = -1$, $\Lambda \geqslant 0$. In case (i) (a), if Λ is positive, show that there is also a solution in which S decreases to a non-zero minimum and thereafter steadily increases towards infinity.

8. If $k = +1$ and $\Lambda = 4/(9D^2)$, show that the radius S of the Friedmann model can first increase from zero to a value $1/\sqrt{\Lambda}$, when the cosmos attains the static Einstein state (section 66) and then, if slightly disturbed from this state, S may either decrease back to zero or continue to increase indefinitely. (This shows that the static Einstein universe is unstable.)

9. A cosmos containing radiation, but no matter, is governed by the equations (68.7) and (68.8). Show that

$$S^2\dot{S}^2 = c^2(D - kS^2 + \tfrac{1}{3}\Lambda S^4)$$

where D is an energy density parameter defined by the equation $3D = \kappa U S^4$.

10. Sketch the graph of $S^2\dot{S}^2$ against S for the universe described in the previous exercise and deduce that all the conclusions listed in exercise 7 are valid if $4/(9D^2)$ is replaced by $3/(4D)$.

11. For the universe described in exercise 9, if $k = 1$, $\Lambda = 3/(4D)$, and $S = 0$ at $t = 0$, prove that at any later time t,

$$S^2 = 2D\{1 - \exp(-ct/\sqrt{D})\}$$

If $S = \sqrt{(2D)}$ at $t = 0$, prove that the universe is static but unstable.

References

Bateman, H. (1952). *Partial Differential Equations of Mathematical Physics*, Cambridge University Press, p. 184.

Bondi, H. (1960). *Cosmology*, Cambridge University Press.

Bondi, H., and Gold, T. (1948). *Mon. Not. Roy. Astr. Soc.*, **108,** p. 252.

Coulson, C. A., and Boyd, T. J. M. (1979). *Electricity*, Longman, p. 309.

Hoyle, F. (1948). *Mon. Not. Roy. Astr. Soc.*, **108,** p. 372.

Rowan-Robinson, M. (1979). *Cosmic Landscape*, Oxford University Press.

Bibliography

Adler, R., Bazin, M., and Shiffer, M. (1965). *Introduction to General Relativity*, McGraw-Hill.
Aharoni, J. (1959). *The Special Theory of Relativity*, Oxford University Press.
Angel, R. B. (1980). *Relativity: The Theory and its Philosophy*, Pergamon.
Berry, M. (1976). *Principles of Cosmology and Gravitation*, Cambridge University Press.
Clark, R. W. (1973). *Einstein: The Life and Times*, Hodder and Stoughton.
Dirac, P. A. M. (1975). *General Theory of Relativity*, Wiley.
Dixon, W. G. (1978). *Special Relativity*, Cambridge University Press.
Einstein, A. (1967). *The Meaning of Relativity*, Chapman and Hall.
Fock, V. (1964). *Theory of Space, Time and Gravitation*, Pergamon.
Frankel, T. (1979). *Gravitational Curvature*, Freeman.
Graves, J. C. (1971). *The Conceptual Foundations of Contemporary Relativity Theory*, M. I. T. Press.
Hawking, S. W., and Ellis, G. F. R. (1973). *The Large Scale Structure of Space–Time*, Cambridge University Press.
Heidmann, J. (1980). *Relativistic Cosmology*, Springer.
Hoffmann, B. (1973). *Albert Einstein, Creator and Rebel*, Hart-Davis and MacGibbon.
Hoyle, F., and Narlikar, J. V. (1974). *Action at a Distance in Physics and Cosmology*, Freeman.
Kilmister, C. W. (1970). *Special Theory of Relativity*, Pergamon.
Kilmister, C. W. (1973). *General Theory of Relativity*, Pergamon.
Lanczos, C. (1974). *The Einstein Decade (1905–1915)*, Elek Science.
Landau, L. D., and Lifshitz, E. M. (1975). *The Classical Theory of Fields*, Pergamon.
Landsberg, P. T., and Evans, D. A. (1977). *Mathematical Cosmology*, Oxford University Press.
Misner, C. W., Thorne, K. S. and Wheeler, J. A. (1973). *Gravitation*, Freeman.
Møller, C. (1952). *Theory of Relativity*, Oxford University Press.
Narlikar, J. (1979). *Lectures on General Relativity and Cosmology*, Macmillan.
Papapetrou, A. (1974). *Lectures on General Relativity*, Reidel.
Petrov, A. Z. (1969). *Einstein Spaces*, Pergamon.
Raychaudhuri, A. K. (1979). *Theoretical Cosmology*, Oxford University Press.
Rindler, W. (1966). *Special Relativity*, Oliver and Boyd.
Rowan-Robinson, M. (1977). *Cosmology*, Oxford University Press.
Schrödinger, E. (1950). *Space–Time Structure*, Cambridge University Press.
Sciama, D. W. (1971). *Modern Cosmology*, Cambridge University Press.
Synge, J. L. (1960). *Relativity–The General Theory*, North-Holland.
Synge, J. L. (1964). *Relativity–The Special Theory*, North-Holland.
Tolman, R. C. (1934). *Relativity, Thermodynamics and Cosmology*, Oxford University Press.
Weinberg, S. (1972). *Gravitation and Cosmology*, Wiley.
Weinberg, S. (1977). *The First Three Minutes*, Deutsch.

Index